International Trade in Hazardous Waste

To our families

To a world of peace and a safer tomorrow

International Trade in Hazardous Waste

D. Kofi Asante-Duah PhD, CE

Environmental Consultant, California, USA and Research Assistant Professor, Center for Environmental Engineering and Science Technologies, University of Massachusetts, Lowell, USA

Imre V. Nagy PhD, DSc

Professor Emeritus, Technical University of Budapest, Hungary

E & FN SPON

An Imprint of Routledge

London and New York

First published 1998 by E & FN Spon, an imprint of Routledge
11 New Fetter Lane, London EC4P 4EE

Simultaneously published in the USA and Canada by Routledge
29 West 35th Street, New York, NY 10001

Typeset in 10/12pt Palatino by Saxon Graphics Ltd, Derby
Printed in Great Britain by St Edmundsbury Press, Bury St Edmunds, Suffolk

British Library Cataloguing in Publication Data
A catalogue record for this book is available from the British Library

ISBN 0 419 21890 4

Contents

Preface

With land space running out for the disposal of toxic wastes and local opposition to such disposal facilities growing, wealthy nations have increasingly been sending wastes abroad to the less affluent nations. The increasing waste quantities, coupled with limited disposal options and tightening environmental regulations have indeed created a difficult situation for waste generators and handlers globally. In particular, as environmental laws in most industrialized nations become increasingly stringent and the cost of waste disposal rises, some waste handlers are turning to cash-poor and easily accessible territories in the developing world. Some developing countries in need of money have continued to accept waste shipments, while others have imposed bans but are unable to monitor them. Such practices are expected to continue and, possibly, even increase in the future. However, any existing waste trade is based largely on uninformed decisions, since the recipient countries in particular tend to be unaware of the risks involved and, furthermore, such countries usually are not adequately compensated. As a reaction to what is essentially an unknown and unregulated situation, the weight of political opinion has shifted to an outright ban on waste trade practices – climaxed by a United Nations (UN) conference vote to ban all waste exports. On the other hand, it is possible to control and regulate the risks involved in such activities so as to make the waste trade more efficient and cost-effective. In a regulated and informed atmosphere, the waste trade could be mutually beneficial to both exporting and importing countries – albeit decisions on human health problems and stability of ecosystems generally should not be dictated by economic considerations.

This book discusses the need for a regulated and informed forum for the international trading activities involving hazardous wastes. It is noted that an outright ban of the waste trade is unworkable and may be an over-reaction to a poorly understood problem. The analysis presented in this book shows that a hazardous waste trade programme can be carefully planned so that informed decisions are made for the benefit of all concerned. Under such circumstances, risks are minimized and net economic benefits are realized in a fair and socially responsible manner. Without advocating widespread commercial or free trade activities in hazardous wastes, the book looks at the pertinent variables and parameters that should be closely reviewed by potential trading partners involved in a waste trade activity, where warranted; the adoption of a regulated and informed strategy will generally allow negotiators to act constructively despite the potential uncertainties inherent in the process.

D. Kofi Asante-Duah PhD, CE
Newport Beach, California, USA
May 1997

Imre V. Nagy PhD, DSc
Budapest, Hungary
May 1997

Acknowledgements

This work has grown out of studies supported under a Robert S. McNamara Fellowships Program administered by the Economic Development Institute of the World Bank, Washington DC. The work was completed while the first author was on appointment as visiting research assistant professor at the University of Waterloo, Ontario, Canada. The views expressed in the book, however, are solely those of the authors and do not necessarily reflect the views of the World Bank or the University of Waterloo.

Our sincere thanks go to the publishing, editorial and production staff at E & FN Spon, London, who helped to bring this book project to a successful completion; and to the Duah family (Abaam, Kade and Nkwantanang), and several friends and colleagues who provided much-needed moral and enthusiastic support throughout preparation of the manuscript. Financial, administrative and material support were also provided by Novelty Environmental Services, Accra, Ghana.

Review comments and suggestions on a preliminary draft were provided by Dr F. Frank Saccomanno (Professor) and Dr John H. Shortreed (Professor and Director, Institute for Risk Research) of the University of Waterloo, Ontario, Canada, and also by Mark Hadley of Environmental Auditors Limited, West Sussex, England. The book also benefited greatly from the review comments of several anonymous individuals, as well as from discussions with a number of professional colleagues. Any shortcomings that remain are, however, the sole responsibility of the authors.

Introduction 1

Wastes are generated at various stages of human activities, and their composition and amounts depend largely on consumption patterns and on industrial and economic structures (OECD, 1994). Waste production is indeed an inevitable characteristic of an industrial or industrializing society. Several views of the inevitable concomitance of waste generation from industrialization have been presented by various authors; in one case summary, Clapham (in Bhatt *et al.*, 1986) portrays hazardous waste management facilities as the kidneys of industrial societies. A responsible system for dealing with hazardous wastes is therefore essential to sustain the modern way of life, in as much as a well-functioning kidney is necessary to rid the human body of certain toxins. The effective management of hazardous wastes, and the associated treatment, storage and disposal facilities (TSDFs), are of major concern not only to the industry producing such material, but also to governments and individual citizens alike due to the nature and potential impacts of such wastes on the environment and public health. Indeed, the actual impacts depend on the waste handling and disposal practices, and are also influenced by the waste composition and origin (OECD, 1994). Consequently, the improvement of hazardous waste management practices is a very important environmental issue for both the industrialized and industrializing countries. But it is also evident that the proper management of hazardous wastes poses great challenges, which have in a way contributed to the extensive interest in global waste trade activities.

Although shipping from the industrialized nations of toxic/hazardous wastes and their dumping onto several developing countries is believed to have been going on for several years, the late 1980s saw increased global media attention to these incidents. The terms **toxic wastes** and **hazardous wastes** are used interchangeably in this book. In general, wastes are considered hazardous if they pose a potential threat to human health or the environment. There are several types and sources of hazardous wastes, most of which are industrial by-products requiring special care during transportation, storage, treatment and/or disposal. In the context discussed here, hazardous wastes are those wastes that generally exhibit the characteristics of ignitability, corrosivity, reactivity, radioactivity and/or toxicity (including carcinogenicity), or any wastes that will meet the World Health Organization (WHO) definition of hazardous waste. The WHO considers hazardous waste as that with physical, chemical or biological characteristics which require special handling and disposal procedures to avoid risk to human health and/or other adverse environmental effects. In fact, a variety of terms are often used by various individual countries to describe these same types of waste.

1.1 THE FATE OF HAZARDOUS WASTES

Hazardous waste management is of great concern in view of the risks associated with such activities. For instance, apart from its immediate and direct health and environmental threats, hazardous waste disposal could lead to the long-term contamination of the ambient air, soils, surface and ground waters, and the food-chain, if disposal facilities are not well designed and maintained, or if remedial actions are not taken in an effective manner.

Increasing concern has emerged during the last decade over the threat posed by the several contaminated sites present in numerous countries, originating from the poor management of hazardous wastes. In fact, until recently only a few countries had established programmes for systematically developing inventories of contaminated sites and their potential risks. This is because building inventories of sites and establishing monitoring networks are expensive, sometimes prohibitively so. This is especially true for situations in Central and Eastern Europe and even more so in the case of developing countries, where there is a need to provide technical and economic assistance to facilitate an assessment of the potential threats from past hazardous waste management practices. In other countries, huge sums of money are currently being spent for the clean-up of sites contaminated from hazardous wastes; typical annual national estimates indicate spendings of about DM 22b. for Germany, HFL 3.5b. for the Netherlands, DKR 400m. for Denmark, and in the USA the estimate is up to about $100b.

Besides the excessive costs involved, there are several other technical and socio-political factors that have made the management of hazardous wastes so problematic. As a consequence, some countries have attempted a 'quick-fix' and easy way out by shipping their wastes abroad.

1.2 BACKGROUND TO THE WASTE TRADE PROBLEM

Yakowitz (1990) enumerates a number of potential stimuli and forces acting to promote the transfrontier movements of hazardous wastes. In particular, as environmental laws in most industrialized nations become increasingly stringent and the cost of waste disposal rises, industry and waste handlers in developed countries find it cheaper to export their wastes to the cash-poor and easily accessible territories in the developing and newly industrializing world, where they pay relatively smaller sums of money towards disposal – far less than what they would spend if they were to dispose of their wastes locally. For instance, a review of the economic aspects of the transport and disposal of hazardous wastes and also other available data on the disposal of hazardous wastes reveal that between 1976 and 1991 the average waste disposal costs in North America may have gone from less than $10 to more than $250 per tonne for landfilling, and from about $50 to over $2600 per tonne for sludge incineration. Depending on a number of external factors, Long (1990) indicates that commercial landfilling in the United States (US) could cost anywhere between $250 and $750 per tonne, whereas incineration of toxic liquids range from $500 to $1000 per tonne; sludge and solid incineration could be in the range of roughly $1000 to $3000 per tonne (Figure 1.1). In Europe, in those countries where solid hazardous wastes can still be legally landfilled, disposal costs are generally lower ($20 per tonne in some places) than in North America; in other countries, these same wastes may be ineligible for landfill, or legal disposal would cost up to $300 per tonne, or generally on average about $70 per tonne (Long, 1990). Indeed, prices vary widely, depending on location, waste type and disposal operation, but this is nonetheless on the rise. Faced with such rising disposal costs and limited disposal options, some industries and waste handlers naturally have taken great interest in waste exports.

As a result of widespread financial crisis, some developing countries have in the past succumbed to pressures to accept wastes for foreign exchange. Recently, it has been observed that toxic wastes have been shipped from certain industrialized countries to some countries in Africa and elsewhere, where it has been costing under $5 per tonne in some cases and no more than about $50 per tonne in most cases to dump such wastes. Consequently, in most of the cases for the waste shipments, the potential savings from wastes exported for land disposal elsewhere could range from $200 to more than $2500 per tonne (on average about $250 per tonne), even if this involves long hauls of several hundreds or thousands of kilometres (Long, 1990). In part this practice takes advantage of an apparently higher tolerance to health/environmental risks in the developing countries, coupled with a critical need for foreign exchange by such nations. In fact, in many of these countries, the immediate need for cash far exceeds long-term concerns about human health and environmental degradation problems – even if purported benefits are only an illusion. Although most of the waste export activities have been in the form of a 'silent' trade, there are noted cases (e.g. CIR and Moyers, 1990; Vallette and Spalding, 1990) where contractual agreements are believed to have

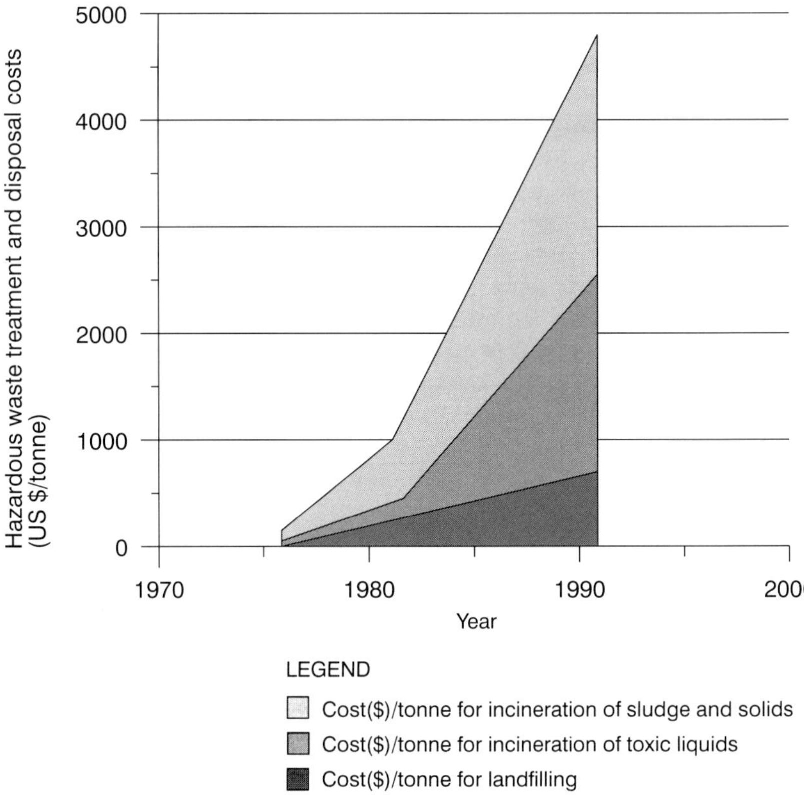

Fig. 1.1 Illustration of the chronological trend in costs associated with the treatment and disposal of hazardous wastes in the United States. *(Source: Approximations, based on information from Lippitt et al., 1986; Long, 1990; and Asante-Duah, 1991.)*

been made with responsible officials or governments of the importing country.

1.3 A PROGNOSIS: CAN THE WASTE TRADE BE CONTROLLED?

In the last two centuries radical changes in production techniques have resulted in an interdependent global economy. Most regions of the world depend on the supply of resources from other areas. Not completely divorced from the resulting forms of commerce is the creation of a marketplace for the by-products of industrialization. In fact, in a way, the hazardous waste trade enables some regions to exceed their local economic capacity via the import of raw materials and energy, and by exporting wastes. Unfortunately, this type of trade activity very often is not based on realistic and fair practices and considerations to all the parties involved; in some regions the trends in hazardous waste trade activities exceed the hazardous waste handling capacity in the importer countries (Rees, 1992). It is noteworthy that hazardous wastes tend to move from countries with more stringent regulation to less regulated countries.

Current patterns in the production and management of hazardous wastes are not likely to change in the short term. The growth of economic activities in Europe, North America and elsewhere is likely to lead to a significant increase in the overall generation of wastes. Consequently, the waste issues will be a dominant concern for both the developed, and especially the developing countries in the coming decades, if waste reduction targets and safe management standards are not achieved in all countries. The waste trade activities are therefore expected to continue and, possibly, even increase in the future. In fact, given the potential for high profits, it is likely that the market for hazardous wastes involving developing countries and newly industrializing countries will continue to be attractive and available. Hence the adoption of an effective international machinery to control and monitor such activities is vital if potential serious long-term negative impacts are to be avoided – or at least minimized.

1.4 COMING TO TERMS WITH THE INTERNATIONAL TRADE IN HAZARDOUS WASTE

Any transboundary movements of wastes across national borders may be viewed as constituting a form of waste trade. Increasing waste quantities from industrial activities, coupled with limited disposal options and tightening environmental regulations, have generally fostered this new kind of trade on the global market-place. This type of trade activity has been taking place between the North and the South and between the East and the West within the international communities. Of particular concerns, however, are the activities that involve the export of toxic wastes from some industrialized nations to certain developing and newly industrializing regions. In any case, the potential risks due to the waste trade activities should be adequately assessed so that sovereign countries can make effective trade decisions regarding development programmes that may include waste trade schemes. Indeed, the potential health risks, environmental impacts and long-term implications arising from a waste trade activity may exceed the short-term economic gains by several orders of magnitude. Furthermore, improper waste management practices may reduce the future development potentials of the country accepting wastes in a trade deal. This, therefore, calls for a careful evaluation of any national policy and development programme that involves waste imports.

This book discusses the need for a regulated and informed forum for the international trading activities involving hazardous wastes. It is noted that a total outright ban of the global waste trade or the transboundary movements of wastes is an unworkable option and may be an overreaction to a poorly understood problem. Clearly, where the need for foreign exchange is great and where certain governments are willing to overlook the risks involved, there will be opportunity for engaging in unethical trade practices. On the other hand, the analysis presented will show that a hazardous waste trade programme can be carefully planned so that informed decisions are made for the benefit of all concerned. Under such circumstances, risks can be minimized and net economic benefits could be realized in a fair and socially responsible manner. But like any other economic activity, the hazardous waste trade must be managed effectively. Effective management requires that the risks associated with individual disposal options are fully understood by all concerned. Where risks exceed established international standards, effective mitigative actions must be taken to reduce such risks to acceptable levels. Furthermore, there should be compensations that reflect the full cost of disposal and treatment programmes borne by the waste recipient country.

Without advocating widespread commercial or free trade activities in hazardous wastes, this book looks at the pertinent variables that should be closely reviewed by potential waste-trading partners involved in a waste trade activity, where warranted. The development and adoption of a regulated and informed strategy will generally allow national or regional policy-makers and negotiators to act constructively, despite the potential uncertainties inherent in the whole process.

This book is designed to provide the necessary directions that may help answer the following question: 'should the toxic/hazardous waste trade be encouraged, and under what terms or circumstances should it be approved?' This question is answered only by adequately assessing the risks, costs and benefits to the trading partners – in particular the waste importing country. The development of a structured conceptual decision framework for waste trade decisions and policies, **where applicable,** will facilitate rational decision-making and negotiations between trading partners involved in a waste trade, **where warranted.** Subsequently, recommendations can be made for actions that policy-makers could take to minimize risks from any decision to trade in wastes.

Legislative-regulatory controls affecting waste management schemes 2

To help abate potential problems to public health and the environment, several items of legislation have been formulated and implemented in most industrialized countries to deal with the regulation of toxic substances present in our modern societies. A general discussion and analysis of hazardous waste management practices in several industrialized countries is presented by Forester and Skinner (1987) in a report of the International Solid Wastes and Public Cleansing Association (ISWA) working group on hazardous wastes. Generally most industrialized countries incorporate a system for 'cradle-to-grave' control in their waste management programmes. The 'cradle-to-grave'-type system monitors and regulates the movement of hazardous materials from manufacture through usage to the ultimate disposal of any associated hazardous wastes. By using some kind of manifest system, this helps minimize abuses and violations of established national or regional control systems associated with hazardous waste movements and management. The manifest system serves as an identification form that accompanies each shipment of wastes; the manifest is signed at each stage of transfer of responsibility (i.e. from the waste generator, to the transporter, to the storage, treatment and/or disposal operator), with each responsible person in the chain of custody keeping a record that is open to scrutiny and inspection by regulatory officials or other appropriate authorities, and also other interested parties. The use of such effective control systems should help minimize the creation of extensive environmental contamination problems. Several of the newly industrializing and other developing countries are also moving towards the establishment of workable environmental regulations, to help protect their often fragile environments.

Invariably there tends to be improved waste management practices in countries or regions where regulatory programmes are well established, whereas those without appropriate regulatory and enforcement programmes become potential targets as dumping grounds in waste trade schemes. The process used in the design of effectual waste management programmes must therefore incorporate several elements of all relevant environmental regulations. In any case, it is noteworthy that variations in national legislations and controls do affect options available for the management of hazardous waste problems in different regions of the world. The need to consult with local environmental regulations when faced with a hazardous waste management problem cannot, therefore, be overemphasized. In fact the establishment of national and/or regional regulatory control programmes within the appropriate legislative-regulatory and enforcement frameworks is a major step in developing effective waste management programmes.

Some of the well-established environmental laws found in the different regions of the world are introduced below, with further examination of related legislations and regulations alluded to in the appropriate sections of subsequent chapters. Depending on the type of programme under evaluation, one or more of the existing regulations may dominate the decision-making process in a waste management programme. It is noteworthy that several of the regulations are subject to changes/amendments. Also, depending on the specific situation or application, a combination of several regulatory specifications may be employed.

2.1 ENVIRONMENTAL LAWS IN NORTH AMERICA

North America, and the US in particular, have over the past several years come up with mountains of environmental legislation. Starting with a National Environmental Protection Act (NEPA) in 1969 that empowered the US Environmental Protection Agency (EPA) and a National Environmental Quality Control Council (NEQCC) to lay the basis for regulations, a steady stream of environmental legislation has followed ever since. Some of the relevant statutes and regulations which, directly or indirectly, may affect waste management decisions, policies and programmes in the US are briefly enumerated below (see also, e.g. USEPA, 1974, 1985; 1987; 1988; 1989).

Clean Air Act. The objective of the Clean Air Act (CAA) of 1970 is to protect and enhance air quality resources, in order to promote and maintain public health and welfare and the productive capacity of the population. This covers all pollutants that may cause significant risks.

Safe Drinking Water Act. The Safe Drinking Water Act (SDWA) was enacted in 1974 in order to ensure that all people served by public water systems would be provided with a supply of high quality water. The SDWA amendments of 1986 established new procedures and deadlines for setting national primary drinking water standards, and established a national monitoring programme for unregulated contaminants, among others. The statute covers public water systems, drinking water regulations and the protection of underground sources of drinking water.

Clean Water Act. The Clean Water Act (CWA) was enacted in 1977, and an amendment to this introduced the Water Quality Act (WQA) of 1987. The objective of the CWA is to restore and maintain the chemical, physical and biological integrity of the nation's waters. This objective is achieved through the control of discharges of pollutants into navigable waters. This control is implemented through the application of federal, state and local discharge standards. The statute covers the limits on waste discharge to navigable waters; standards for discharge of toxic pollutants; and the prohibition on discharge of oil or hazardous substances into navigable waters. Another closely associated legislation, the Federal Water Pollution Control Act (FWPCA), deals solely with the regulation of effluent and water quality standards.

Resource Conservation and Recovery Act. The Resource Conservation and Recovery Act (RCRA) was enacted in 1976 (as an amendment to the Solid Waste Disposal Act of 1965, later amended in 1970 by the Resource Recovery Act) to regulate the management of hazardous wastes, to ensure the safe disposal of wastes, and to provide for resource recovery from the environment by controlling hazardous wastes 'from cradle to grave.' This is a federal law that establishes a regulatory system to track hazardous substances from the time of generation to disposal. The law requires safe and secure procedures to be used in treating, transporting, sorting and disposing of hazardous substances. Basically RCRA regulates hazardous waste generation, storage, transportation, treatment and disposal. The 1984 Hazardous and Solid Waste Amendments (HSWA) to RCRA, Subtitle C, covers a management system that regulates hazardous wastes from the time they are generated until the ultimate disposal (the so-called 'cradle-to-grave' system). Thus, under RCRA, a hazardous waste management programme is based on a 'cradle-to-grave' concept, that allows all hazardous wastes to be traced and equitably accounted for.

Comprehensive Environmental Response, Compensation and Liability Act. The Comprehensive Environmental Response, Compensation and Liability Act of 1980 (CERCLA or 'Superfund') establishes a broad authority to deal with releases or threats of releases of hazardous substances, pollutants or contaminants from vessels, containments or facilities. This legislation deals with the remediation of hazardous waste sites, by providing for the clean-up of inactive and abandoned hazardous waste sites. The objective is to provide a mechanism for the federal government to respond to uncontrolled releases of hazardous substances into the environment. The statute covers reporting requirements for past and present owners/operators of hazardous waste facilities; and the liability issues for owners/operators for cost of removal or remedial action and damages, in case of release or threat of release of hazardous wastes. The Superfund Amendments and Reauthorization Act of 1986 (SARA) strengthens and expands the clean-up programme under CERCLA; focuses on the need for emergency preparedness and community right-to-know; and changes the tax structure for financing the Hazardous Substance Response Trust Fund established under CERCLA to pay for the clean-up of abandoned and uncontrolled hazardous waste sites.

Toxic Substances Control Act. The Toxic Substances Control Act (TSCA) of 1976 provides for a wide range of risk management actions to accommodate the variety of risk-benefit situations confronting the US EPA. The risk management decisions under TSCA would consider not only the risk factors (such as probability and severity of effects), but also non-risk factors (such as potential and actual benefits derived from use of the material and availability of alternative substances). Broadly TSCA regulates the manufacture, use and disposal of chemical substances. It authorizes the US EPA to establish regulations pertaining to the testing of chemical substances and mixtures; pre-manufacture notification for new chemicals or significant new uses of existing substances; control of chemical substances or mixtures that pose an imminent hazard; and record-keeping and reporting requirements.

Federal Insecticide, Fungicide and Rodenticide Act. The Federal Insecticide, Fungicide and Rodenticide Act (FIFRA) deals with published procedures for the disposal and storage of excess pesticides and pesticide containers. The US EPA has also promulgated tolerance levels for pesticides and pesticide residues in or on raw agricultural commodities under the authority of the Federal Food, Drug, and Cosmetic Act (40 CFR Part 180). FIFRA does indeed provide the US EPA with broad authorities to regulate all pesticides.

Endangered Species Act. The Endangered Species Act (ESA) of 1973 (reauthorized in 1988) provides a means for conserving various species of fish, wildlife and plants that are threatened with extinction. The ESA considers an endangered species as that which is in danger of extinction in all or a significant portion of its range; a threatened species is that which is likely to become an endangered species in the near future. Also, the ESA provides for the designation of critical habitats (i.e. specific areas within the geographical area occupied by the endangered or threatened species) on which are found those physical or biological features essential to the conservation of the species in question.

Hazardous Materials Transport Act. The Hazardous Materials Transport Act (HMTA) provides legislation that deals with the regulation of transport of hazardous materials.

Fish and Wildlife Conservation Act. The Fish and Wildlife Conservation Act of 1980 requires States to identify significant habitats and develop conservation plans for these areas.

Marine Mammal Protection Act. The Marine Mammal Protection Act of 1972 protects all marine mammals – some of which are endangered species.

Migratory Bird Treaty Act. The Migratory Bird Treaty Act of 1972 implements many treaties involving migratory birds, thus giving protection to most species of native birds in the US.

Wild and Scenic Rivers Act. The Wild and Scenic Rivers Act of 1972 preserves select rivers declared as possessing outstanding remarkable scenic, recreational, geologic, fish and wildlife, historic, cultural or other similar values.

Comparable laws also exist in Canada; however, the implementation and enforcement of their environmental regulations (with a few exceptions) have been principally the responsibilities of the provinces. In many ways, the Canadian system is less formal than the US system, but not any less stringent.

Other laws and regulations exist in the US and Canada that also work towards preventing or limiting the potential impacts of hazardous waste problems. Also, laws similar to those annotated above, together with similar legal provisions and regulations can be found in the legislative requirements for several other developed countries, as reflected in the brief discussions in sections 2.2 and 2.3.

2.2 THE GENERAL CHARACTER OF HAZARDOUS WASTE LEGISLATION IN EUROPE

For the most part, the European countries have consistently approached the issues of waste management legislation from very different angles. For example, whereas there are very tight legal controls in Germany, hazardous waste disposal practices in Italy very often take place with little regard to any legal requirements. Denmark, Sweden and Finland have opted for centralized multifunctional facilities: flow control is applied to ensure that these facilities receive the appropriate waste types and quantities. The United Kingdom (UK) and France take the view that very careful land use planning can help to ensure that waste disposal practices will not create future threats to the environment. Ireland and Luxembourg typically export their hazardous wastes for disposal elsewhere. Notwithstanding the differences in waste management legislation in Europe, there is a general tendency for the development of hazardous waste minimization policies in most of the countries in the region. In fact, several European countries have already adopted, or are developing regulatory instruments that will directly affect the amount of hazardous wastes produced.

Typically the costs of treatment and disposal have been recognized as an important factor in promoting waste minimization. European governments influence treatment and disposal costs primarily by setting higher regulatory standards for treatment and disposal facilities (i.e. requiring specific wastes to be managed using particular methods, technologies or facilities) and by making it more difficult to site and develop new waste management facilities. As a result, several major European chemical manufacturers have established corporate waste minimization policies that are implemented through a variety of programmes (such as waste budgets, centralized environmental review of new investments, and research and development of cleaner production technologies). Ultimately, the expected goals are an increased production efficiency and competitiveness, lower waste management costs, and enhanced corporate reputation with the consumers of their products and/or the general public.

2.2.1 ENVIRONMENTAL LAWS WITHIN THE EUROPEAN UNION

The European Union (EU) was created by the Treaty of Maastricht, and came into existence in November 1993 as an umbrella organization for the European Community (EC), which was created in 1957 by the Treaty of Rome. The EU presently consists of the following 15 Member nations: Austria, Belgium, Denmark, Finland, France, Germany, Greece, Italy, Luxembourg, the Netherlands, Portugal, Republic of Ireland, Spain, Sweden and the UK. Fundamentally the EC is a structural member of the EU.

The EU directly affects Member States through various legal instruments and several guiding principles. The key principles that shape environmental policies of the EU's Member States include the 'polluter pays' principle; the need to avoid environmental damage at the source; and the need to take the environmental effects of energy and other EU policies into consideration.

The two most common forms of EU legislation are regulations and directives; regulations become law throughout the EU as of their effective date, generally enforceable in each Member State, whereas directives are generally not necessarily enforceable in Member States but are meant to set out goals for the Member States to achieve through national legislation. In fact, of the five major legislative acts enumerated in Article 189 of the EC Economic Treaty (i.e. Treaty Establishing the European Economic Community, Rome, 25 March 1957) – namely regulations, directives, decisions, recommendations and opinions – the directive seems to play a unique role that affords the European parliament the ability to respect indigenous national legislative traditions and philosophies while guiding the EU as a whole towards a single common goal. The directive sets forth an objective and provides that each national government achieve this objective by the national means considered best able to accomplish the goal embodied in the directive. It was designed to respect indigenous legislative traditions while providing a means for unification of goals.

Specifically the EU directives play an important role in environmental regulations within the Member States. For instance, the 1975 framework (Council Directive 75/442/EEC on Waste) established the general obligations of Member States for waste management, to ensure that wastes are disposed of without endangering human health or harming the environment. Toxic and dangerous wastes were covered under a 1978 directive (Council Directive 78/319/EEC on Toxic and Dangerous Waste) that required these materials to be properly stored, treated and disposed of in authorized facilities governed by proper regulatory authorities; this directive has more recently been replaced with the Directive on Hazardous Wastes (Council Directive 91/689/EEC), adopted under the provisions of the framework Directive on Waste, and this new directive took effect at the start of 1994. The environmental directives are designed to establish uniform results with respect to environmental integrity throughout the Member States. While striving towards a common goal of environmental consistency, the environmental directive respects national policy for the management, control and regulation of environmental pollutants.

A key element of the relevant EU environmental directives is the recent concept of using the 'Best Available Techniques Not Entailing Excessive Costs' (BATNEEC) to prevent or reduce pollution and its effects. BATNEEC is interpreted as the technology for which operating experience had adequately demonstrated it to be the best commercially available to minimize releases, providing it had been proven to be economically viable.

Overall, the countries within the EU have aggressively worked on producing consistent environmental regulatory programmes for their Member nations, but also recognize the sovereign rights of the individual countries. As part of this process, the European Environment Agency was created in 1990 by a Community Regulation (Council Regulation 1210/90/EEC on the Establishment of the European Environment Agency and the European Environment Information and Observation Network), with the main purposes of collecting and disseminating information, rather than to have any enforcement authority. In fact, it is apparent that a vast and comprehensive legislative arsenal has been created within the EU and therefore in each of the Member States, albeit the implementation of these legislations is yet to achieve the complete anticipated results.

Some of the apparently unique legislative measures and relevant statutes and regulations which, directly or indirectly, may affect waste management decisions, policies and programmes within specific/selected EU nations are briefly annotated below – recognizing that similar or comparable frameworks exist in the other Member nations. Further discussions of several aspects of environmental regulatory programmes in the region appear in the legislative control literature elsewhere (e.g. Garbutt, 1995a,b; Lister, 1996).

UK. Waste management has always been an issue visited by governments and policy-makers, past and present alike. For instance, the Mediaeval Orders called for the removal of refuse on pain of forfeits (e.g. the Henry IV Order of 1407). Before the 19th century, however, pollution problems arising from the disposal of waste or refuse were frequently resolved through private litigation rather than by public prosecution under the Sanitary Acts. During the current era, the Public Health Act of 1936 enacted that any accumulation or deposition of refuse prejudicial to health or a nuisance is a Statutory Nuisance. The regulatory authorities have the power to serve Abatement Notices and to seek public prosecution; this power will remain extant, even after the full implementation of the Deposit of Poisonous Waste Act (DPWA) of 1972 and the Control of Pollution Act (CoPA) in 1974.

Early in 1972 the DPWA had been issued, penalizing the deposit on land of poisonous, noxious or polluting wastes so as to give rise to an environmental hazard, and to make offenders criminally and civilly liable for any resultant damage. The UK Control of Pollution Act of 1974 allows the government to restrict the production, importation, sale or use of chemical substances in the UK; the Act controls the disposal of wastes on land by means of site licensing. Subsequently, the 1980 Control of Pollution (Special Waste) Regulations, which also repealed the DPWA, was established to ensure pre-notification for a limited range of the most hazardous types of wastes; to keep a 'cradle-to-grave' record of each disposal of special wastes; and to keep long-term records of locations of special waste disposal landfill sites.

The Environmental Protection Act (EPA 90) 1990 marked an important milestone in the development of the legislative philosophy and framework in the UK. The Act is seen as being of major importance, since it largely established Britain's strategy for pollution control and waste management for the foreseeable future (D. Slater, in Hester and Harrison, 1995).

Most recently the Environment Act 1995 sought to develop a new breed of environmental regulations in the UK. The significance of this Act includes (among other issues/things) a reorganization and concentration/integration of environmental regulatory functions, and its adoption is complemented by pivotal reliance on cost-benefit and risk assessment techniques as mechanisms for its implementation (Jewell and Steele, 1996). The Act brings together in a single body, the existing environmental protection and pollution control functions of a number of regulators.

The Netherlands. The Netherlands has adopted a general ban on land disposal of hazardous wastes. When exemptions from this ban are granted, the government may require the recipient of the exemption to conduct research on alternative technologies or management methods aimed at preventing potential needs for future land disposal. The ban itself acts, to some extent, as an incentive for waste minimization, although the relatively easy access to waste export to other countries for treatment or disposal has so far limited the impact of such a ban.

France. In France, the main goals of environmental policy for the management of special industrial wastes are: to develop an effective national treatment and disposal infrastructure, reduce the amount of wastes generated at source, and clean up sites contaminated by hazardous wastes. The French Ministry of the Environment's plan to attain these goals has been via providing clear definitions of the different classes of wastes, and imposing regulatory and financial constraints on waste generators.

Measures provided for in recent legislation include restrictions on the types of wastes that can be landfilled, and a tax on the offsite treatment and disposal of special industrial wastes, that will be used to fund the clean-up of abandoned contaminated sites. The special industrial wastes will include medical wastes and hazardous domestic wastes. The relevant decree lists 51 groups of compounds, as well as 14 chemical and biological properties that are likely to render a waste hazardous.

The French government encourages waste avoidance and reduction by including waste production and management practices among the factors that are evaluated when the government has to make a decision on whether or not to authorize the operation of new or modified industrial facilities. An impact study is required prior to any authorization. However, the potential impact of these requirements on the reduction of hazardous wastes is probably not being realized for the time being, because of an insufficient number of enforcement personnel and inadequate technical expertise among regulators and industry concerning alternative technologies and management methods.

From 2002 disposal by landfill in France is only to be permitted for wastes or treatment residues for which there is no commercially available recycling or treatment technology, and that require disposal by landfill. The present legislation requires that, by 1998, all special industrial wastes must be stabilized prior to disposal.

Federal Republic of Germany. In Germany industrial waste is managed under the Federal Waste Disposal Act, which defines a category of hazardous waste as 'special', i.e. requiring particular/specific handling procedures. If a facility does not generate or handle 'special' wastes by definition, the agency may authorize the facility to handle the waste by itself, with only minimal regulatory requirements.

A regulatory instrument adopted by the German parliament in 1986 provides an accelerated waste reduction policy, and authorizes the government to regulate the use, collection and disposal of products likely to cause environmental problems when discarded as waste. The regulation is coordinated with other efforts such as public financial support for developing and demonstrating new waste avoidance technologies, increasing the technical competence of regulators, and discouraging reliance on treatment and disposal as methods of choice for managing hazardous wastes.

Overall, the federal system for the management of industrial wastes consists of a basic legislative framework that delegates significant authority to the state for the management and implementation of industrial waste regulations.

2.2.2 ENVIRONMENTAL LAWS AND CONTROLS FOR THE FORMER SOVIET BLOC COUNTRIES OF EASTERN AND CENTRAL EUROPE

Since the collapse of the Soviet Union's 'iron curtain', environmental problems in former Soviet bloc countries of Central and Eastern Europe have gradually been coming to light, and fundamental changes are already taking place in these new societies. The result is the development of systematic environmental control programmes, most of which follow the styles employed by the industrialized countries in Europe, North America and elsewhere in the world.

2.3 ENVIRONMENTAL POLICIES AND REGULATIONS IN ASIA, THE PACIFIC REGIONS AND OTHER PARTS OF THE WORLD

The legislative measures and relevant statutes and regulations which, directly or indirectly, affect waste management decisions, policies and programmes generally tend to be similar in their degree of sophistication according to the level of technological advancement. This means that developed countries in Asia and elsewhere will have regulatory systems comparable to those found in Western Europe or North America (as exemplified by the brief references given below for Japan and Australia). On the other hand, many developing countries tend to lack laws and/or enforcement tools that regulate environmental problems in an effective manner. In fact, information and waste management regulations have been almost non-existent and/or non-enforceable in most developing countries. This is because most developing countries lack the laws and governmental institutions to deal with environmental regulatory issues. Many governments, however, recognize the need to protect their respective nation's health and environment, and are therefore seriously developing legislation that will deal with the unsurmountable waste management problems.

Japan In Japan the Waste Management Law of 1970 requires anyone undertaking the collection, transport or disposal of industrial wastes to have obtained a permit; waste disposal sites are also required to be designed and operated in a manner that prevents trespassing, and the disposal sites must be isolated from both surface and ground water resources, and measures taken to prevent leakage. In addition, the 1973 Chemical Substances Control Act requires a manufacturer or importer of a new chemical (i.e. one that is not enumerated in a 1973 pre-listed compilation) to provide the government with all available information concerning such a chemical; the substance can then be evaluated and classified as being dangerous or safe.

As part of the nation's overall waste management programme, Japan has put in place laws and regulations, as well as enforcement mechanisms, that are comparable to those found in most other industrialized countries.

Australia Environmental protection is administered in different ways in different parts of Australia. Recent environmental legislations in the region include the Clean Air Act 1961, the Clean Waters Act 1970, the State Pollution Control Commission's Enabling Act 1970 and the Commonwealth Environmental Protection Act 1974. In fact, much of the pollution legislation currently in force in Australia dates back to the 1960s and 1970s, with public health as its main focus (UNICRI, 1993).

Responsibilities for national environmental legislation and controls lie with the Australian Commonwealth government. In general, regulations differ only slightly from one state to another, although New South Wales (NSW) probably has the most comprehensive and stringent environmental legislation yet. In any case, observers have noted that, whereas legislation is generally uniform, monitoring and enforcement tend to differ depending on such factors as community attitudes, population pressure, land-use conflicts, etc.

2.4 NEW TRENDS AND CONTEMPORARY ISSUES IN THE LEGISLATIVE-REGULATORY CONTROL SYSTEMS

In the past some polluting behaviours were seen as normal and inevitable consequences of industrialization and national development. As a result, environmental protection in many countries was based on reliance on a non-punitive, conciliatory style of securing compliance with environmental laws and regulations (UNICRI, 1993). Increasing damage to the environment has, however, changed perceptions of acts of pollution, especially in the industrialized countries.

There is now a growing tendency for the developed countries to move steadily towards criminal sanctions as a means of punishment and deterrence in the field of environmental offences. In Canada, for instance, the Law Reform Commission has advocated the use of criminal prosecution and strict criminal enforcement as a deterrent (Rebovich, 1992). In Europe the offences regarding hazardous waste disposal will be treated more stringently than before. For example, the proposed amendments to a German law on environmental crime would mean significant changes in the existing system of penalties. The bill proposes to increase to ten years, the maximum prison sentence for particularly severe environmental offences, and to introduce generally higher fines for environmental crimes. It also includes an expansion of the regulations for 'active repentance', under which the severity of a penalty can be reduced or the charges dropped altogether, if the responsible party voluntarily remedies the environmental damage done or has made serious attempts to do so. The bill contains stricter penalties for non-compliance with hazardous waste disposal regulations. It will also impose stricter punishment for 'waste tourism' and the illegal export and import of hazardous wastes.

In fact, increasing public concern about the problems and potentially dangerous situations associated with waste management, together with the legal provisions of various legislative instruments and regulatory programmes, have all compelled both industry and governmental authorities to carefully formulate responsible waste management programmes. These programmes include techniques and strategies needed to provide good waste management methods and technologies, and the development of cost-effective corrective action programmes that will ensure public safety, as well as protect human health, the environment, and public and private properties.

2.5 A PARADIGM OF INTERNATIONAL ENVIRONMENTAL LAW

Transboundary pollution problems have been around for a long time and this scenario is not about to change. Hence, to be truly effective, environmental protection usually must have an international dimension to it. Solutions to cross-boundary environmental issues will therefore tend to require international cooperation.

Since solutions to global environmental problems generally require increased international cooperation – both regionally and globally – it should be expected that international environmental law will subjugate national sovereignty to this end. In this vein, it is apparent that, the legislation and implementation of most international environmental laws have the tendency to neglect the social factors and the differing economic development status of states, whereas the objective reality is that any global issue is an inseparable and organic whole (L.P. Cheng, in Weiss, 1992). On this basis it becomes necessary for the international community, in dealing with global environmental issues, to consider the objective conditions of those states that are less developed economically and technologically as an almost separate 'entity' from the industrialized nations. This strategy will, hopefully, help alleviate part of the grim situation associated with the legislation and implementation of international environmental protection plans or laws by several nations/states. In fact, the differing economic and political situations of states may have a dramatic impact on the implementation of international environmental laws in different countries and this should be recognized if successful stories are to be told about existing and new international environmental laws.

Global dimensions of waste disposal practices

3

Economic development programmes in several countries not only depend on the trade in natural resources from other regions, but also the activities involved usually result in the production of hazardous wastes – which often could have global consequences. Even so, currently there is no comprehensive inventory of waste production, composition and disposal practices for the global community. This is due largely to the fact that only a few countries have a comprehensive system in place for reporting and/or documenting such information. In Europe, for instance, only a few countries have adopted regulatory requirements for industries to report on the amount of hazardous wastes generated. But, the most comprehensive assessment yet of Europe's environment has confirmed the existence of serious environmental degradation in many countries in the region, especially in states of the former Soviet Union (Burke, 1996; Stanners and Bourdeau, 1995). Economic, social, technical and institutional issues will clearly affect how a specific region or country chooses specific policies with respect to waste management schemes.

The quantities and types of waste produced in several countries are often reported only on a voluntary basis, and not necessarily as part of any systematic monitoring activity. In addition, waste statistics across countries are often not comparable due to divergence in definitions and differences in the waste classification systems from one country to another. For example, the generic terms 'municipal waste', 'industrial waste' and 'hazardous waste' may include different waste streams in each national context and/or in different time-scales. In fact, historical data for hazardous waste information are seldom consistent with more recent information. More recent data are based on ad hoc surveys or estimates; estimates based on production and consumption trends usually have a wide margin of error. Also, a considerable portion of hazardous wastes still escapes control and, therefore, is often not recorded at all in official statistics. In particular, some materials destined for recovery operations are often not recorded as wastes.

In any case, there are no simple answers to the diverse situations found in global waste management. In fact, the quantities of various types of waste materials which are generated and that have to be disposed of are huge and still growing, particularly in the developed countries. Furthermore, costs of remedial action are typically high, and the threat of adverse health and environmental effects is never completely removed. Many countries have indeed experienced adverse consequences resulting from the improper management of potentially hazardous wastes, even if different in nature and dimension.

3.1 WASTE MANAGEMENT PRACTISED IN INDUSTRIALIZED NATIONS

There is increasing public concern about the problems and potentially dangerous situations associated with waste management practices. Such concerns, together with the legal provisions of various legislative instruments and regulatory programmes, have all compelled both industry and governmental authorities to carefully formulate waste management plans. These plans include techniques and strategies needed to provide good waste management methods and technologies, and the development of cost-effective corrective action programmes that will ensure public safety, as well as protect human health, the environment and public and private properties.

In general many developed countries tend to have large industrial bases that are major sources of hazardous waste streams. A market survey will reflect national trends in spending on pollution control to be on the increase globally; for most industrialized countries, such spending is about 1 to 1.5% of the gross domestic product (GDP). The amount spent to control pollution is generally a reflection of the volume of industrial activity, and the nature of environmental policies and regulations in the different countries.

Overall, waste management as practised by Member countries of the Organization for Economic Cooperation and Development (OECD) generally serves as a model for much of the rest of the world. Pursuant to a convention signed in Paris in December 1960, and which came into force in September 1961, the original Member countries of the OECD included: Austria, Belgium, Canada, Denmark, France, Germany, Greece, Iceland, Ireland, Italy, Luxembourg, the Netherlands, Norway, Portugal, Spain, Sweden, Switzerland, Turkey, the UK and the US; the following countries became Members subsequently through accession: Japan in April 1964, Finland in January 1969, Australia in June 1971, New Zealand in

Table 3.1 Early 1990s hazardous waste statistics showing the relative contributions from selected OECD countries. (Source: OECD, 1994.)

Country	Total annual production ('000 tonnes)	Relative contribution to OECD totals (%)
USA	180 000	75
Canada	6 080	2.6
Germany	6 000	2.5
France	3 958	1.7
Italy	3 246	1.4
UK	2 940	1.2
Spain	1 708	0.72
Portugal	1 043	0.44
Netherlands	1 040	0.44
Switzerland	736	0.31
Japan	666	0.28
Austria	620	0.26
Sweden	500	0.21
Greece	450	0.19
Australia	316	0.13
Finland	314	0.13
Turkey	300	0.13
Norway	200	0.08
New Zealand	110	0.05
Denmark	106	0.04
Iceland	5	0.002
All OECD	237 388	

May 1973, and most recently Mexico in May 1994 (OECD, 1994). In 1974 the OECD Environment Committee, which guides all work involving environmental matters on behalf of its Member countries, created the Waste Management Policy Group. Composed of high level governmental officials, this group is charged with the responsibility for working to achieve environmentally sound waste management in their respective countries. The group is primarily concerned with evaluating and developing international policies to promote appropriate waste management as an integral part of sustainable growth; the process involved considers technical, economic and social issues and seeks to harmonize the results of these studies into coherent and implementable international policy actions.

3.1.1 SOURCES OF WASTES AND HAZARDOUS WASTE STREAMS IN DEVELOPED COUNTRIES

Wastes are generated from several activities, including industrial, agricultural, military, commercial and even domestic operations. According to the OECD, the total amount of waste produced in 1990 in OECD countries was about 9 billion tonnes, of which 425 million tonnes was municipal wastes and 1.5 billion tonnes was industrial wastes (including more than 230 million tonnes of hazardous wastes); over 6.5 billion tonnes of the wastes were residues from energy production, agriculture, mining, demolition and sewage sludge (OECD, 1991, 1994). For the most part, these figures have been growing over the years, but the relative contributions from the various OECD countries remains very much the same (Table 3.1).

Overall, economic growth in the developed countries has been driven to a considerable extent by progress achieved by the chemical industry. At present more than 10 million chemical compounds have been formulated, and up to 300 new chemicals apparently enter the market each year, with the current world production of chemicals set at about 400 million tonnes (Lonngren, 1992). Production of chlorinated organic compounds, for example, have increased dramatically in the last two decades because of their use in metal cleaning and degreasing, dry-cleaning, and paint stripping. Within the EU countries alone, it is estimated that between 600 000 and 800 000 tonnes of chlorinated solvents were used in 1989; of that, methylene chloride accounted for the largest portion of approximately 160 000 tonnes, followed by trichloroethylene (TCE), perchloroethylene (PCE), and 1,1,1-trichloroethane (1,1,1-TCA) at approximately 130 000 tonnes each (CEC, 1992). Residual wastes from the use of these chlorinated solvents in the EU countries is estimated to be about 200 000 tonnes, with only approximately 90 000 tonnes being recycled.

Several other types of wastes do indeed contribute to the total volume of waste streams present in the industrialized world. Among others, used tyres are the most important source of waste rubber products in terms of volume; the total amount of used tyres present in the EU alone has been estimated at over 1.9 million tonnes per annum, with about 46% disposed of by landfilling, approximately 31% transferred for recovery of materials or energy, and about 23% reused as retreaded tyres. Wastes from the packaging of consumer products are yet another significant source of both industrial and municipal wastes; OECD Europe alone produces about 45 million tonnes of packaging waste per year, with a recovery rate between 10 and 15%. Also of great concern is the management of radioactive wastes within Europe; for example, radioactive waste generation is estimated to consist of approximately 160 000 tonnes per annum of heavy metal, plus some 3400 tonnes of spent nuclear fuel and about 300 tonnes of highly radioactive vitrified waste from spent fuel reprocessing (CEC, 1993). In fact, it is projected that radioactive waste production will increase significantly by the turn of the century due to new wastes resulting from the decommissioning of obsolete nuclear plants especially from the former Soviet bloc countries.

Table 3.2 Major industries and manufacturers contributing to hazardous waste streams in industrialized countries.

Industry/Manufacturer
• Aerospace
• Ammunitions
• Automobile
• Batteries
• Beverages
• Chemical production
• Computer manufacture
• Electronics and electrical
• Electroplating and metal finishing
• Explosives manufacture
• Food and dairy products
• Herbicides, insecticides and pesticides
• Ink formulation
• Inorganic pigments
• Iron and steel
• Leather tanning and finishing
• Medical care facilities
• Metal smelting and refining
• Mineral exploration and mining
• Paint products
• Perfumes and cosmetics
• Petroleum products
• Pharmaceutical products
• Photographic materials
• Printing and publishing
• Pulp and paper mills
• Rubber products, plastic materials and synthetics
• Shipbuilding
• Soap and detergent manufacture
• Textile products
• Wood processing and preservation

In summary, Table 3.2 offers a listing of typical industries generating large volumes of hazardous wastes in most industrialized countries. These industries generate several waste types such as organic waste sludges and still bottoms containing chlorinated solvents, metals, oils, etc.; oil and grease contaminated with polychlorinated biphenyls (PCBs), polychlorinated terphenyls (PCTs), polybrominated biphenyls (PBBs), polyaromatic hydrocarbons (PAHs), metals, etc.; heavy metal solutions of arsenic, cadmium, chromium, lead, mercury, etc.; PCB wastes; pesticide and herbicide wastes; anion complexes containing cadmium, copper, nickel, zinc, etc.; paint and organic residuals; and several miscellaneous chemicals and by-products that need special handling/management.

3.1.2 WASTE MANAGEMENT STRATEGIES WITHIN THE OECD COUNTRIES

The OECD countries, which account for only about 16% of the world's population, produce approximately 93% of the world's total industrial wastes, with Europe being responsible for approximately 38% of the total (OECD, 1991). Large amounts of these wastes and certain chemical compounds that are now banned or have restricted use in the OECD countries (e.g. DDT and other pesticides) are sometimes exported to other non-OECD countries, potentially causing human health and environmental problems in the destination countries. Notwithstanding the shortcomings, the OECD probably has the best waste management strategies worthy of emulation by the global community; the OECD waste control options, ranked in order of desirability from the point of view of environmental impeccability, are enumerated as follows (OECD, 1993).

1. Reduction of waste generation by use of efficient manufacturing processes, reduction of disposable material in consumer goods and/or increase of products durability.
2. Use of efficient methods to control effluents, including the separation of reusable components at their source.
3. Direct reuse of waste products whenever possible, return of effluents to the production process (as in the case for steel-making or cement kiln operations), or exchange of material which is considered as waste from one process but may be a feedstock for another process.
4. Transformation or other physical or chemical treatment, in order to recycle usable materials from waste (e.g. burning of household wastes to recover energy, magnetic separation of ferrous scrap from waste and used to prepare ferrous products, reclamation of non-ferrous metals from mixed industrial wastes by thermal processes, re-refining of waste lubricating oils, or distillation and regeneration of spent solvents).
5. Destruction of the waste by physico-chemical treatment or incineration (e.g. neutralization by mixing alkaline and acid wastes, or burning of pumpable liquid waste or solid wastes).
6. Permanent storage of the waste in or on land, or use of the sea as a disposal area, or incineration-at-sea.

Where the waste management strategy involves transboundary movements, OECD countries have opted to use a classification system for wastes destined for recovery operations. This distinguishes wastes in terms of the 'green', 'amber' and 'red' lists – depending on the level of control applicable to their transfrontier movements. These lists have been published in EC Council Regulation 93/259/EEC on the supervision and control of shipments of wastes.

To facilitate implementability of several aspects of the waste control options in Europe, for example, there has been an initiative to create a general nomenclature for waste and to develop a European Waste Catalogue (EWC). The EWC has been approved (EC Decision 94/3/EC), in compliance with EC Council Directive on Waste 75/442/EEC as amended by EC Directive 91/156/EEC, which requires a common reference list of waste to be established across the EU Member States. Its implementation should provide a common basis for cross-referencing national lists, and facilitate the implementation of EU waste management policies and also should allow EU Member States to harmonize waste control systems and reporting on wastes. Comparable strategies to facilitate the implementation of the OECD-type waste control options are being considered by several other industrialized countries, and even by some of the developing economies.

3.1.3 ISSUES AFFECTING MONITORING, CONTROL AND WASTE DISPOSAL PROGRAMMES

A number of developed countries have established national systems for the monitoring and control of hazardous wastes. The central goal of such systems is to ensure that discarded substances which society generally view as potentially harmful to human health and/or to the environment are managed in such a manner as to minimize the possibility for adverse effects to occur. The following specific characteristics have been identified as being common, and indeed important, to all monitoring and control systems that are designed for hazardous waste management programmes (OECD, 1993).

- Definition of 'wastes' and 'disposal', and the setting of a boundary between 'wastes' destined for disposal and 'materials' destined for recycling, resource recovery, reuse or reclamation.
- Designation of a list of wastes which will constitute 'hazardous wastes'.
- Compilation of a list of probable sources of hazardous wastes.
- Assessment of the quantity of each type of waste to be controlled.
- Provision of treatment, storage and disposal facilities (TSDFs) to match the quantities of each type of hazardous wastes that are produced.
- Implementation of a monitoring scheme which will afford the competent authorities the means to track the wastes from 'cradle-to-grave', and perhaps beyond.
- Means to respond to any emergencies caused by accident, system failure or discovery of abandoned hazardous wastes which may pose threats to humans and/or the environment.
- A regimen for assigning responsibility and liability with respect to the management of hazardous wastes.
- Adequate resources for setting compliance standards with the regimens selected to control hazardous wastes, and for enforcing the rules in case of non-compliance.

- A long-term approach showing the benefits to be expected from any proposed policy changes.

A number of factors usually come to bear on how individual countries choose to emphasize the various elements; these include costs, geography, industrial mix, public awareness, legislative mandate and indeed several other case-specific variables.

3.1.4 A SYNOPSIS OF WASTE MANAGEMENT METHODS AND TECHNOLOGIES

Many technologies and processes are currently available for the management of hazardous wastes. However, any one particular technology or process is usually not appropriate for all waste types and/or under all varying conditions existing at specific geographical locations. An in-place or onsite technology or process will generally be given preference as a method of choice in the development of management options for hazardous waste streams. The efficiency and effectiveness of an in-place process or technique depends on both contaminant concentration levels and quantity of each chemical constituent in the waste stream. In most situations waste treatment processes can generate organic and inorganic compound emissions; under these sets of conditions, volatile organic chemical (VOC) emissions and fugitive dust generation have particularly great potential to contaminate the air, and are therefore of major concern in the choice of treatment technologies.

This section gives a brief general overview of the more commonly used waste management technologies and processes that find extensive application in most industrialized countries. It does not include the detailed design elements of the various technologies, nor does it discuss new and developing technologies. A more detailed description of the processes, equipment and controls, as well as the detailed design elements of various technologies, can be found elsewhere in the literature of waste management technologies (e.g., Acar *et al.*, 1993; Anderson, 1993–95; ARB 1991; Avogadro and Ragaini, 1994; Cairney, 1987, 1993; Charbeneau *et al.*, 1992; Haas and Vamos, 1995; Jolley and Wang, 1993; NRC, 1994; Nyer, 1992, 1993; OBG, 1988; Pratt, 1993; Reed *et al.*, 1995; Sims, 1990; Sims *et al.*, 1986; Tedder and Pohland, 1990–95; USEPA, 1984, 1985, 1988, 1989, 1990; Vandegrift *et al.*, 1992; Vidic and Pohland, 1995; Wang and Wang, 1992; Wise and Trantolo, 1994).

3.1.4.1 Treatment processes

Waste management techniques and processes finding widespread applications in several developed countries may be classified into the following general and broad categories annotated below.

Physical treatment techniques and processes. Some chemical constituents can be treated through separation and purification processes. These types of techniques generally do not alter the chemical composition of the components of waste streams and, therefore, are referred to as physical treatments. Physical treatments include technologies such as filtration, centrifugation, floatation, distillation, evaporation, solvent extraction, reverse osmosis, ion exchange, adsorption, decantation and constituent immobilization by solidification; these are processes generally applied without the aid of chemical or biological agents. For instance, activated carbon adsorption technology is based on the principle that certain organic constituents preferentially adsorb to organic carbon. It should be noted that physical processes are rarely used as the final treatment and disposal option for any given waste stream. For example, in the clean-up of contaminated groundwater and other diluted liquid wastes, mobile carbon regenerators have commonly been used to sorb the hazardous constituents onto granular activated carbon; the spent carbon must then be shipped to furnaces for regeneration by combustion.

Chemical treatment methods. Some types of waste streams can be either separated or rendered less hazardous through chemical treatment processes. For instance, by using a specific solvent, chlorine atom(s) can be removed from chlorinated hazardous materials, and toxic compounds are converted to less toxic, more water-soluble compounds; the reaction products are more easily removed from soil and more easily treated. Chemical treatment technologies often applied to hazardous waste streams include neutralization, coagulation-precipitation, and oxidation-reduction (redox). Typically, the addition of chemicals followed by a sequence of mixing, flocculation, coagulation and settlement is useful either before or following other treatment processes. In leachate treatment, for example, the use of hydrated lime, quick lime, sodium hydroxide, magnesium hydroxide, alum, ferric chloride, ferric sulphate or polymeric coagulant aids may be very effective in achieving the following results: reduction in suspended solid concentrations so as to minimize problems in subsequent biological processes; reduced problems with blockage of pipes and valves and, in activated carbon columns, the precipitation of iron, manganese, calcium carbonate and heavy metals in order to protect physical plant and prevent toxicity and inorganic solids build-up in biological processes; removal of turbidity and colour from effluents; and the partial removal of organic loading. The oxidation-reduction type reactions are used selectively to reduce the toxicity of certain hazardous waste constituents. However, chemical detoxification processes have generally been waste-specific, which has precluded its application to the complex chemical mixtures encountered in hazardous wastes.

Biological treatment processes. Biological treatment has become a viable and cost-effective alternative technology for treating a wide variety of chemical constituents (such as petroleum hydrocarbons, aromatic hydrocarbons, chlorinated solvents and pesticides). By using micro-organisms (natural or engineered) to degrade chemical constituents, biological processes transform hazardous/toxic materials into non-toxic elements such as water, carbon dioxide and other innocuous products. The transformation or degradation of organic chemicals by microbes constitutes biological treatment. Biological treatment generally has no effect on dissolved inorganic substances, even though significant levels of some inorganic chemicals may inhibit biological activity, or may even kill the microbes. Biological treatment is also highly sensitive to changes in the organic composition of the waste being treated. Biodegradation is less efficient at low substrate concentrations, because if the concentration is too low the compound may not be metabolized by a microbial population which may favour another substrate that is available in higher concentrations; on the other hand, very high substrate concentrations may be toxic to the microbial community.

Thermal treatment alternatives. Thermal treatment technologies employ heat to destroy or change the contaminant of concern; the most common thermal alternatives are incineration, catalytic conversion and in-situ vitrification. Thermal degradation is applicable to waste streams that contain significant concentrations of organic compounds, and can be implemented through different types of incineration (controlled, high-temperature burning in the presence of oxygen) and pyrolysis (chemical degradation due to elevated temperatures, without the presence of oxygen).

3.1.4.2 Treatment methods

Many developed countries use several miscellaneous methods and practices that may be classified as variations or combinations of physical, chemical, biological and/or thermal treatment techniques for hazardous waste management. Some of those finding relatively widespread applications are introduced below.

Incineration. Incineration is a thermal treatment process by which the waste streams are exposed to excessive heat in one of several different types of available incinerators; the different incinerator types are designed and built for different purposes. The process involves the thermal destruction of contaminants by burning; depending on the intensity of the heat, the contaminants of concern are volatilized and/or destroyed during the incineration process.

Two types of proven incineration processes – liquid injection and rotary kilns – dominate today's market. Liquid injection units atomize wastes into the combustion chambers, normally requiring auxiliary fuels to maintain the temperature at about 800°C; the combustion is completed in a secondary chamber operated at about 1000°C. Liquid injection units are effective on large volume wastes, particularly those with a high calorific value. Rotary kilns are especially useful for organic contaminants in soils or other solid residues; temperatures in the kiln are normally in the range of 700 to 900°C. Volatile and partially combusted organics are carried in the gas stream to an after-burner where auxiliary fuel raises the temperature to about 900 to 1000°C in order to complete the destruction process. The ash remaining from incineration of hazardous wastes is generally disposed of in landfills. Incineration of high calorific value wastes may indeed be regarded as a form of recycling, if the heat generated is utilized for some beneficial uses. In general different incinerator types will typically be selected for different waste streams. For example:

- for the incineration of soils or sludge contaminated with organic hazardous constituents (such as PCBs, pesticides, dioxins and creosote sludges), the Electric Infrared Furnace can be used, where infrared energy is provided by electrically powered silicon carbide rods;
- in an advanced fluidized bed incinerator, such as the Circulating Bed Combustor, where high velocity air suspends a bed of solid particles in the combustion chamber, there is no afterburner. The suspended bed materials that leave the combustion chamber are recovered in a cyclone and recirculated through the furnace. It is useful for burning materials such as high sulfur-coal, peat, wood, waste oils and municipal waste;
- the newly developed Pyroplasma System prolyzes liquid organic wastes using a thermal plasma field which produces a thermal plasma with temperatures of more than 5000°C. It is potentially capable of destroying such chemicals as methyl ethyl ketone, methanol, ethanol, carbon tetrachloride and PCBs;
- the Oxygen Enriched Burner uses pure oxygen in combination with air and natural gas to destroy liquid hazardous wastes. The use of oxygen allows higher temperatures which improves the kinetics of waste destruction, allows higher waste throughput and reduces stack gas volume. The burner should be effective in destroying wastes which require high temperatures for destruction (e.g. halogenated organics), or have low heating values (small amounts of organic constituents in an inert base).

In fact, there is a continuing progress in thermal and combustion technology, but emissions of products of incomplete combustion (PICs) remain a concern, because some thermal breakdown products can be as toxic as (if not more toxic than) the hazardous constituents from which they were derived. The current level of knowledge is limited in the extent to which PICs in incinerator emissions pose a hazard to human health and/or the environment.

Catalytic conversion. Catalytic conversion is an incineration process that uses a catalyst to reduce the usually high incineration temperature requirements, or for special treatment applications. Catalytic units can be built for specialized contaminants, but the cost of such units tends to be rather high.

In-situ vitrification. In-situ vitrification (ISV) is a thermal treatment process that converts waste materials in soils into a chemically inert and stable glass and crystalline product, by using electrical heat to melt the soil. It may be used to provide solution to mixed wastes (including organic, inorganic and radioactive wastes) in soils.

Chemical fixation (or stabilization). This is a technique to chemically fix or modify the chemical structure of hazardous materials, by applying specific reagents. The fixation process consists of immobilizing the chemicals in the waste stream in-place, thereby preventing any migration into various environmental media. The waste materials are blended with precise amounts of reagent(s) to stabilize and/or encapsulate chemical constituents which are then stockpiled and allowed to cure; often, the treated material is rendered non-hazardous.

Chemical fixation can be applied to many organic- and metal-bearing waste streams. In general the chemical fixation processes are waste-specific, and are limited to the types of wastes that can be economically treated. Current techniques include cement-based and lime-based, thermoplastics, organic polymer encapsulation, self-cementing and glassification processes.

Encapsulation/landfilling. This is the physical isolation and containment of the waste stream, that often includes a physical barrier installation at the sides of the contamination, since its purpose is to leave the contaminant safely in place at a chosen location. The contaminated material is covered with low-permeability layers of synthetic textiles or clay; the cap is designed to limit infiltration of precipitation and thus prevent the migration of contaminants away from the site and into groundwater.

Typical problems which have been experienced in the land disposal practice include: the poor quality control during installation of liner and cap systems, resulting in seam leaks in synthetic liners and channelling through clay liners; incompatibility of liner materials with wastes; puncture of liners during operation; and subsidence of wastes causing cracks in the cap.

Landfarming (land treatment). This is a treatment process by which waste materials are spread over an area to enhance naturally occurring processes such as volatilization, aeration, biodegradation and photolysis. Standard earth-moving equipment is used to prepare the landfarm area and for applying the material for treatment. The treatment area usually consists of a low-permeability liner material to prevent or minimize leaching, and is also surrounded by berms to control potential run-off processes and erosion from the soil materials during rain storm events. Treatment requirements include periodic applications of nutrients, moisture control and discing for oxygen exposure; biological treatments use bacteria and other micro-organisms to degrade contaminants.

Vapour extraction systems. When waste materials have already escaped into contaminated soils or groundwater, vapour extraction systems (VESs) may be applied to the removal of VOCs from the contaminated media. The common applications include in-situ soil venting or in-situ air stripping (which uses soil vapour extraction systems); air or steam stripping; and air sparging.

- *In-situ soil venting*, or *soil vapour extraction* (*SVE*), is a technique which uses soil aeration to treat sub-surface zones of contaminated soil for VOCs. It is the process of removing and venting VOCs from the vadose or unsaturated zone of contaminated soils.
- *Air or steam stripping* is a physical treatment process, that involves the physical removal of dissolved-phase contamination from a water stream. This method is frequently used to remove VOCs from the groundwater after it is pumped to the surface.
- *Air sparging* is the highly controlled injection of air into the contaminant plume in the saturated zone. Used in conjunction with soil vapour extraction, it is emerging as an effective treatment technology for soils and groundwater contaminated with VOCs.

The VES may vary considerably in size and design, depending on site-specific requirements and/or conditions. A typical, well-designed VES consists of a series of extraction/injection wells connected to a common manifold and to a positive displacement air blower, among other surface equipment.

Waste recycling/materials reuse. Material recyling and reuse is an important part of general waste management practices. For instance, the recovery of waste oils, solvents and waste heat from incinerators is a common practice in several countries. Also, operations exist in a number of countries to recycle heavy metals from various sources, such as silver from photo-finishing operations; lead from lead-acid batteries; mercury from batteries and broken thermometers; heavy metals from metal finishing wastes; etc. Indeed, waste recycling is an integral part of many industrial processes. Furthermore, waste exchange schemes exist in some countries, to promote the use of one company's by-product or waste as another's raw material.

3.1.4.3 Treatment options and waste management strategies

The above list of treatment methods is by no means complete. In general there are several proven waste management technologies and processes

Table 3.3 Summary of selected waste treatment methods and technologies commonly applied to waste management programmes.

Nature of waste/matrix of concern	Treatment option	Basic technology and process	Scope of potential applications	Limitations/ comments	Estimated treatment costs (per volume of impacted matrix)
Solids	Asphalt batching	Asphalt incorporation, involving the incorporation of petroleum-contaminated soils into hot asphalt mixes as a partial substitute for stone aggregate.	• Economical for larger volumes of contaminated materials • Dependent on climate; most asphalt plants do not operate during cold weather.	• May be unsuitable for clays. • May require offsite transportation.	\$40–\$105 per cubic metre (\$30–\$80 per cubic yard)
	Bioremediation	Natural or enhanced biodegradation, involving a process to degrade organic compounds into innocuous materials. Needs shallow groundwater (<15 meters [or 50 feet]) or an underlying impermeable silt or clay layer.	• Most cost-effective for large volumes of contaminated material. • Most applicable when contamination extends into groundwater, and is of sufficient volume or depth below surface. • Minimal disruption to site operations, for in-situ bioremediation.	• Labour-intensive; requires considerable maintenance. • possibility for contaminant migration. • Loss of efficiency in soils containing certain chemicals and low pH.	\$15–\$195 per cubic metre (\$10–\$150 per cubic yard)
	Encapsulation	In-situ containment and isolation process that comprises isolating contaminated soils from	• Used only to prevent contaminant migration. • Applicable to most chemicals, provided	• Long-term monitoring required. • Does not destroy contaminants; only	\$40–\$195 per cubic metre (\$30–\$150 per cubic yard)

Table 3.3 *(continued)*

Nature of waste/matrix of concern	Treatment option	Basic technology and process	Scope of potential applications	Limitations/ comments	Estimated treatment costs (per volume of impacted matrix)
		the surrounding environment by use of clay caps, liners, slurry walls, grout curtains, etc.	such a compound does not attack containment materials. • Open areas preferred. • Example application involves capping landfills to prevent leaching by recharge water.	prevents migration. • Solution may not be permanent.	
	Chemical fixation	Solidification/stabalization treatment process, involving the addition of materials to decrease the mobility of the original waste constituent.	• Applicable to a wide variety of waste materials.	• Long-term monitoring required. • Future land-use may be restricted.	$130–$390 per cubic metre ($100–$300 per cubic yard)
	Excavation/landfill disposal	Process involves the removal of contaminated soils.	• Best for removal and disposal of limited volumes of 'hot spot' materials from a vast area that may otherwise be 'clean'. • Common practice for shallow, highly contaminated soils.	• Potential for long-term liabilities. • Long-term monitoring required. • Trend is towards increasing disposal costs.	$65–$390 per cubic metre ($50–$300 per cubic yard)

Incineration	Thermal treatment	•Economy of scale for large volumes. •Applicable to a broad range of organic compounds.	•Generally high costs. •Extensive permit requirements for on site treatment. •Potential disposal problems for residual materials; residual ash may require further treatment.	\$50–\$390 per cubic metre (\$40–\$300 per cubic yard)
In-situ soil flushing (in-situ soil leaching)	In-situ soil leaching process that involves injecting or flushing in-place soils with water to leach chemicals in soils into groundwater; surfactants may be added to facilitate the flushing process.	•Applicable to both organics and inorganics, but to different degrees. •Most applicable when contamination extends to groundwater table, requiring use in conjunction with extraction and treatment systems. •Site hydrogeology has strong influence.	•Leachate collection required. •Long-term monitoring required. •Potential problems with leaching fluid used. •Less feasible for complex mixtures of waste types. •Costs depend on site characteristics, contaminant consituents, and clean-up levels.	\$25–\$160 per cubic metre (\$20–\$120 per cubic yard)
Landfarming	Land treatment process, by which affected soils are are removed and spread over a treatment area in a layer so as to enhance naturally occuring degradational processes.	•Best suited for lighter organic compounds. •Warmer temperatures are conducive to faster degradation rates, since temperature influences the rate of degradation.	•Emissions control is difficult. •Low clean-up levels may not be practical. •Requires relatively large treatment areas. •Certain chemicals may be toxic to native microbes that could be	\$15–\$50 per cubic metre (\$10–\$40 per cubic yard)

Table 3.3 *(continued)*

Nature of waste/matrix of concern	Treatment option	Basic technology and process	Scope of potential applications	Limitations/ comments	Estimated treatment costs (per volume of impacted matrix)
				facilitating degradation process.	
	Passive remediation	A 'no action' option that relies entirely on several natural processes to destroy the contaminants of concern.	•Site-specific; greater depth to groundwater, presence of aquitard, and low infiltration may minimize migration to groundwater. •Temperature may affect volatilization and natural degradation.	•Long-term monitoring required. •Long-term liabilities possible. •Low restoration levels may not be possible. •Effectiveness may be influenced by soil condition. •Possible future land-use restrictions.	$0 (for remediation) – low cost (for monitoring programmes)
	Vapour extraction systems	In-situ volatilization, involving the removal of volatile organic compounds from sub-surface soils by mechanically drawing air through the soil matrix.	•Applicable to VOCs. •Can be applied to wide area extents. •Minimal disruption to site operation; conducive to both developed and undeveloped sites. •Most effective at higher concentrations. •Generally low costs.	•Venting and emissions difficult to control for very shallow areas. •Not effective below the groundwater table. •Performance affected by soil conditions; less-effective and/or longer time frame in	$15–$185 per cubic metre ($10–$140 per cubic yard)

				fine-grained soils. • May not work for semi-VOCs.	
Liquids	Air stripping	Aeration process, effective for removal of VOCs.	• Simple technology, usable in conjunction with other methods. • Low maintenance costs.	• Problems with air emissions; collection and treatment systems required for off-gas releases.	\$15–\$185 per cubic metre (\$10–\$140 per cubic yard)
	Carbon adsorption	Activated carbon adsorption technology, based on the principle that certain organic contaminants preferentially adsorb onto organic carbon.	• Simple technology, usable in conjunction with other methods. • Attractive as point-of-use treatment method. • More appropriate for aquifer restoration. • Applicable to a broad range of (organic and inorganic) contaminants.	• Generates 'spent carbon' as a waste by-product. • High maintenance costs; carbon requires frequent regeneration. • Not considered cost-effective nor efficient for aquifer restoration.	\$15–\$25 per cubic metre (\$10–\$20 per cubic yard)
	Passive remediation	A 'no action' option that relies entirely on several natural processes to destroy the contaminants of concern.	• Several physical and chemical variables (e.g. temperature and alkalinity) may affect natural degradation and other attenuation processes.	• Long-term monitoring required. • Long-term liabilities possible. • Low restoration levels may not be possible.	\$0 (for remediation) – low cost (for monitoring programmes)

used amongst the developed countries, but no single technology can universally be applied with equal success to all waste types and in all geographical regions. In fact, treatment processes can be, and usually are, combined into process trains for more effective removal of hazardous materials from waste streams. Table 3.3 presents a summary of the treatment options commonly employed in a variety of waste management programmes.

It is apparent that several countries have been relying predominantly on landfill for disposing of their waste. In the past, most landfills were rudimentary waste dumps without the necessary protective requirements. Currently, however, measures to control landfill releases are specified in environmental regulations adopted by most countries, in particular the developed countries. The second most common method of waste management in several OECD countries is by incineration, which is considered most suitable to reduce waste volumes (by as much as 80%), and also for energy recovery. However, several concerns exist about the health effects of toxic substances that are released during incineration. Incineration typically produces vast quantities of fly and bottom ashes, which contain hazardous metals (such as cadmium, chromium, lead and mercury) and also highly toxic organic compounds (e.g. dioxins and furans).

Finally, it is noteworthy that recycling in a number of OECD countries increased between 1975 and 1985 but seems to have declined in much of Europe since 1985 because of the increasing waste production trends and the difficulties in finding appropriate markets for recovered materials. Apparently, virgin materials are generally more economical choices than secondary or recycled ones. Consequently, the export of hazardous wastes still remains a more competitive option than waste minimization and/or recycling. The difficulties with recycling are universal, but the motivation for both waste minimization and recycling remains on the growth path globally.

3.1.5 WHERE DO WE GO FROM HERE?

In view of the large quantities of wastes that have always been generated by many industries, and because of the apparent crisis with managing such wastes, waste recycling and reuse, waste exchange, and waste minimization are becoming more prominent in the general waste management practices of several developed/industrialized countries. Figure 3.1 illustrates the basic components of a typical waste management programme for an industrial facility in an industrialized country. The general trend is to favour onsite waste management, with more emphasis for the future directed at waste minimization. Onsite waste minimization to be applied to waste management programmes will comprise:

- waste recycling (i.e. the recovery of materials used or produced by a process for separate use or direct reuse in-house);

- materials recovery (i.e. the processing of waste streams to recover materials which can be used as feedstock for conversion to another product);
- waste reuse (i.e. the direct reuse of the wastes as they are, without further processing); and
- reduction at source (that may comprise process modifications and/or the introduction of good housekeeping methods).

All these aspects of the waste management programme will likely dominate the scene in industrial waste management planning of the future; however, waste reduction at source in particular will gain more prominence in the long term. Reduction at source will be accomplished through process modifications and through efficient or good housekeeping. The process modification aspect of waste reduction at source may generally consist of the following typical elements in particular:

- modification of the process pathway
- improved control measures
- equipment modification
- changes in operational setting/environment
- increased automation
- product substitution.

Good and efficient housekeeping can be attained by implementing the following typical tasks:

- material handling improvements
- spill and leak monitoring, collection and prevention
- preventative maintenance
- inventory control
- waste stream segregation.

When every effort has been made at minimizing wastes, then whatever wastes are generated can be permitted for long-term onsite storage; or for the permanent onsite disposal in landfills, surface impoundments, etc.; or for an end-of-pipe treatment onsite to reduce waste quantity or contaminant concentrations prior to offsite disposal. When considered for offsite management, wastes should first be considered for local waste exchange programmes or other treatment, storage and disposal options locally before any consideration is given to waste exports.

3.2 WASTE MANAGEMENT PRACTICES IN DEVELOPING NATIONS AND COUNTRIES WITH ECONOMIES IN TRANSITION

In order to break the shackles of age-long poverty and to catch up in economic status with other nations in the industrialized world, most developing countries have accepted the challenges of development strategies based on rapid industrialization. In fact, most developing countries appear to be moving towards open-market economic growth through

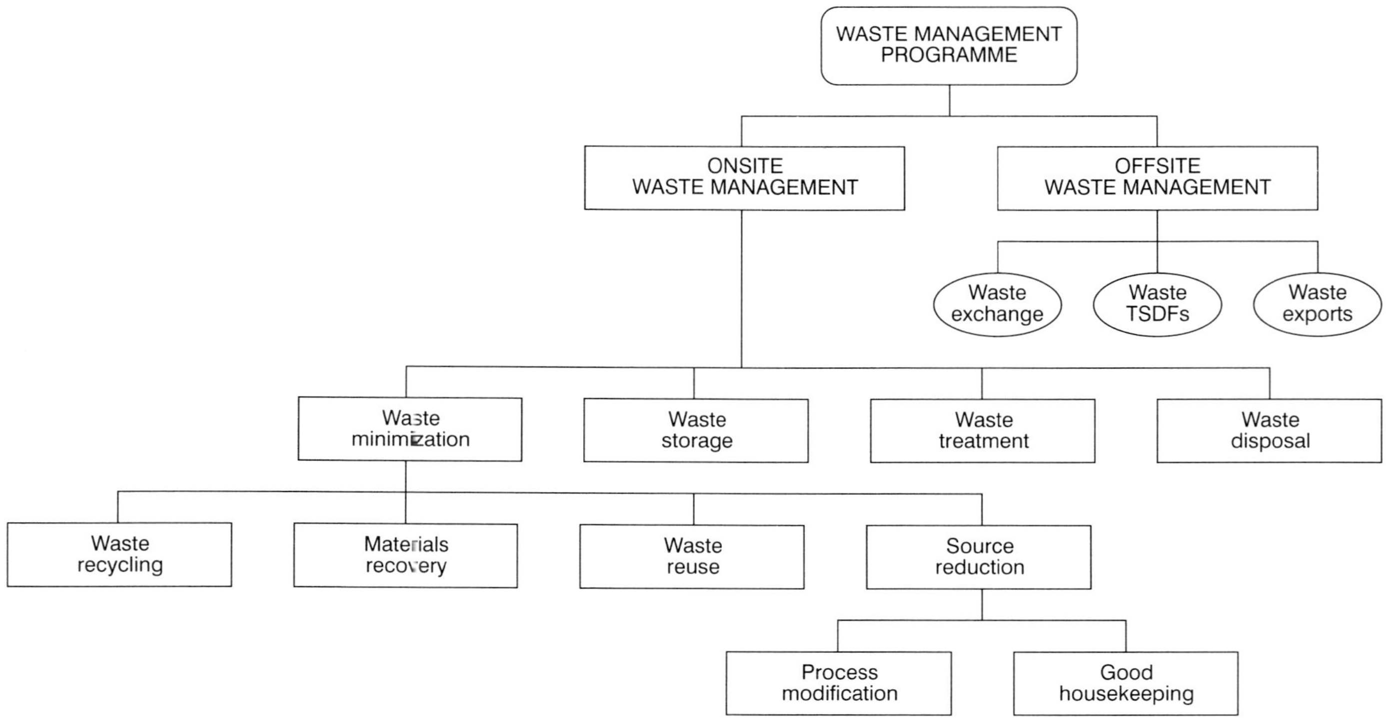

Fig. 3.1 Basic components of a typical waste management programme.

'Western-style' industrialization. The economic development efforts of the developing countries have been directed towards an increasing interest in industrialization, since this is generally regarded as the panacea for underdevelopment, and this is believed to be necessary for socio-economic development in such nations. There is the general belief that industrialization would eventually lead to an enhanced economy, an increase in a country's gross national product (GNP), and a concomitant enhancement of the general quality of life in a region. But accompanying this effort of industrialization is the generation of several types of wastes that mimic what has happened in the industrialized countries, except that in the case of the developing countries there is no machinery in place to effectively manage such wastes. Generally there is the absence of specific national legislation and enforcement policies governing the generation, storage, treatment and disposal of hazardous wastes in most developing countries. Populations are therefore ultimately put at risk, and the benefits derived from the projected economic growth are – possibly – latently destroyed, or at least partially offset by the health and environmental hazards created by these very economic activities.

3.2.1 SOURCES OF WASTES AND HAZARDOUS WASTE STREAMS IN DEVELOPING COUNTRIES

The more visible forms of wastes identifiable in most developing countries are municipal solid wastes. For instance, in Nigeria, heaps of refuse and garbage have been reported as a common sight in most state capitals and other urban areas, especially in the period of extreme economic rejuvenation during the oil boom (FMHE, 1982; Holmes, 1984). Apart from the potential health threats from such open dumps, this disposal practice – which is suspected to be widespread in several developing and newly industrializing countries – is not even aesthetically acceptable, because most of the dumps are located in the heart of central business districts and/or in residential areas.

Of even greater concern is the generation of hazardous wastes from industrial and related activities that take place within developing nations and countries with economies in transition. The primary sources of hazardous wastes present in most industrializing and/or developing countries include:

- wastes generated locally by small quantity generators and households;
- wastes generated locally by medium-sized private enterprises and industries;
- wastes generated locally by major large-scale firms that are state-owned, foreign-owned or joint-ventures; and
- wastes imported into the country through some form of trade deal, or other bilateral or multilateral agreements.

Small quantity hazardous waste generators constitute an important part of most developing nations' total volume of hazardous wastes. Table 3.4 lists the major small quantity generators found in several developing countries; most of these generators – especially photographic works, dry-cleaners, and automotive service and repair shops – are very often located close to or within residential developments/settings. The common waste types generated by these small quantity generators do indeed contribute significantly to the overall hazardous waste generation in many developing countries. On the other hand, wastes originating from small quantity generators and domestic sources are probably the most difficult to control. Hazardous wastes in developing countries, therefore, tend to present potentially greater problems than similar wastes produced in the regulated world of industrialized nations.

The major industrial sectors that remain potential sources of waste generation in most developing and/or newly industrializing countries, are shown in Table 3.5. In general the manufacturing, mining, transportation, agricultural and service industries all generate wastes that may impact human health and the environment. However, the chemicals and allied products are generally seen as the major sources of industrial hazardous waste generation. Table 3.6 summarizes the typical hazardous waste streams potentially generated in several developing countries by the various sectors of the economies; these waste streams have the potential to enter the various environmental compartments and cause adverse effects on human and ecological receptors in the region.

Lastly, hazardous waste problems in developing countries may also arise from hazardous products produced in developed countries and sent to developing countries for use in such industries as agriculture, raw

Table 3.4 Typical small quantity waste generators in developing countries.

Waste generator
• Ceramics
• Distilleries
• Furniture works
• Laundries/dry-cleaners
• Metal manufacturing
• Pesticide application services
• Photography
• Printing press
• Soap manufacture
• Vehicle maintenance

Table 3.5 Categories of major industries and manufacturers in developing countries.

Industry/manufacturer
• Batteries
• Breweries
• Distilleries
• Dyes and pigments
• Fertilizers
• Food processing and dairy products
• Fruit and vegetable packaging
• Mining and metallurgy
• Motor assembly/repairs
• Motor vehicle body manufacture
• Oil milling
• Paper and paper products
• Perfumes and cosmetics
• Pesticides
• Petroleum
• Pharmaceutical and chemical products
• Printing
• Shipbuilding
• Soaps and detergents
• Structural and metal products
• Sugar and confectionery
• Tanneries and leather goods
• Textiles
• Tobacco
• Wood and furniture

materials extractions, metal-processing and mining operations, energy, and also from hazardous wastes exported from industrialized countries and dumped onto the developing nations. In a classic example, when the use of PCBs in electrical equipment was outlawed in most developed countries, electrical transformers and capacitors which could not be used due to their PCB content were sold to Brazil in very large quantities. As a consequence, there are millions of electrical products containing PCB still being used in Brazil and the associated wastes are generally disposed of without adequate levels of care. This typical example should constitute an important warning to developing countries wishing to achieve sound development through industrialization (Nogueira, 1988). Additionally an important source of hazardous wastes introduced into the environment in most developing nations are due to hazardous materials typically found in commercial products in various sectors of the society; some typical specific chemicals appearing in a variety of commercial products commonly imported and used in several developing countries are identified in Table 3.7.

Table 3.6 Examples of potentially hazardous waste streams from various industrial sectors and other sources in developing countries.

Sector/source	Typical hazardous waste stream
Agricultural and food production	Acids and alkalis; fertilizers (e.g. nitrates); herbicides (e.g. dioxins); insecticides; unused pesticides (e.g. aldicarb, aldrin, DDT, dieldrin, parathion, toxaphene)
Airports	Hydraulic fluids; oils
Auto/vehicle servicing	Acids and alkalis; heavy metals;lead-acid batteries (e.g. cadmium, lead, nickel); solvents; waste oils
Chemical/pharmaceuticals	Acids and alkalis; biocide wastes; cyanide wastes; heavy metals (e.g. arsenic, mercury); infectious and laboratory wastes; organic residues; PCBs; solvents
Domestic	Acids and alkalis; dry-cell batteries (e.g. cadmium, mercury, zinc); heavy metals; insecticides; solvents (e.g. ethanol, kerosene)
Dry-cleaning/laundries	Detergents (e.g. boron, phosphates); dry-cleaning filtration residues; halogenated solvents
Educational/research institutions	Acids and alkalis; ignitable wastes; reactives (e.g. chromic acid, cyanides; hypochlorites, organic peroxides, perchlorates, sulphides); solvents
Electrical transformers	Polychlorinated biphenyls (PCBs)
Equipment repair	Acids and alkalis; ignitable wastes; solvents
Leather tanning	Inorganics (e.g. chromium, lead); solvents
Machinery manufacturing	Acids and alkalis; cyanide wastes; heavy metals (e.g. cadmium, lead); oils; solvents
Medical/health services	Laboratory wastes; pathogenic/infectious wastes; radionuclides; solvents

Metal treating/manufacture	Acids and alkalis; cyanide wastes; heavy metals (e.g. antimony, arsenic, cadmium, cobalt); ignitable wastes; reactives; solvents (e.g. toluene, xylenes)
Military training grounds	Heavy metals
Mineral processing/extraction	High-volume/low-hazard wastes (e.g. mine tailings); red muds
Motor freight/railroad terminals	Acids and alkalis; heavy metals; ignitable wastes (e.g. acetone; benzene; methanol); lead-acid batteries; solvents
Paint manufacture	Heavy metals (e.g. antimony, cadmium, chromium); PCBs; solvents; toxic pigments (e.g. chromium oxide)
Paper manufacture/printing	Acids and alkalis; dyes; heavy metals (e.g. chromium, lead); inks; paints and resins; solvents
Petrochemical industry/petrol stations	Benzo-a-pyrene (BaP); hydrocarbons; oily wastes; lead; phenols; spent catalysts
Photofinishing/photographic industry	Acids; silver; solvents
Plastic materials and synthetics	Heavy metals (e.g. antimony, cadmium, copper, mercury); organic solvents
Shipyards and repair shops	Heavy metals (e.g. arsenic, mercury, tin); solvents
Textile processing	Dyestuff; heavy metals and compounds (e.g. antimony, arsenic, cadmium, chromium, mercury, lead, nickel); halogenated solvents; mineral acids; PCBs
Timber/wood preserving industry	Heavy metals (e.g. arsenic); non-halogenated solvents; oily wastes; preserving agents (e.g. creosote, chromated copper arsenate, pentachlorophenol)

Table 3.7 Examples of hazardous chemicals potentially appearing in commercial products with possible widespread use in developing countries.

Chemical	Typical commercial product
Asbestos	Insulations/reinforcement (in various rubber, plastic or cement products)
Benzene	Dry-cleaning fluids; fumigants; insecticides; petrol; motor oil; paint remover; solvents
Cadmium	Household batteries; motor oils; rubber from automobile tyres
Carbon tetrachloride	Degreasers; fumigants; refrigerants; solvents
Chloroform	Anaesthetics; fumigants, insecticides; pharmaceuticals; solvents
Dichlorodiphenyl trichloroethane (DDT)	Pesticides
1,2 Dichloroethane	Degreasers; finish and paint removers; fumigants; petrol; soaps; solvents
Ethylbenzene	Petrol
Ethylene dibromide	Fumigants; petrol; solvents
Formaldehyde	Binding agents/glue; disinfectants; embalming agents; preservatives
Lead	Petrol (tetraethyl lead); lead-acid storage batteries; paints
Lindane	Paints; prescriptions for human lice and scabies infestations
Mercury	Household (mercury) batteries; paints; thermometers
Methylene chloride	Degreasers; dewaxers; fumigants; furniture refinishers; hairspray; paints; paint removers; shoe cleaner/polish; solvents
Pentachlorophenois (PCPs)	Wood preservatives
Polychlorinated biphenyls (PCBs)	Electrical transformers
Tetrachloroethylene (TCE)	Degreasers; dry-cleaning solvents; paint removers
1,1,1-Trichloroethane	Degreasers; furniture polish, shoe polish; paint removers; pesticides; solvents
Trichloroethylene	Adhesives; degreasers; dry-cleaning fluids; dyes; paints; fumigants; pharmaceuticals; shoe cleaner/polish; solvents
Vinyl chloride	Adhesives for plastics

3.2.2 HOW ARE INTERNALLY GENERATED WASTES MANAGED?

Presently it is almost impossible to obtain any reliable information regarding chemical waste generation and disposal practices, and the associated health effects in most developing countries. However, those is a general impression that pollution from hazardous wastes is occuring in the countries in these regions, due to uncontrolled discharges from industrial facilities and agricultural practices; the consequences are those of environmental degradation, especially the degradation of water resources. Also household hazardous wastes (comprising solvents, paints and several others) are typically not segregated and are therefore co-disposed with other garbage at municipal disposal sites. Open dumping and burning of wastes, which is prevalent in several developing countries, would usually leave potentially dangerous by-products with serious implications to public health and the environment – including air pollution and generation of dioxin-based ashes. Ashes from the burn are sometimes (consciously or unconsciously) spread out on farms, thus propagating the contamination problem. Furthermore, semi-controlled tipping or the creation and use of crude sanitary landfills – such as simply filling up depressions and valleys – that do not include methods to control leachate generation and contaminant migration results in the pollution of potential sources of potable water. Burrow-pits and erosion gullies are also used for disposal in several locations. Yet in other cases, some farmers apparently stockpile solid waste for conversion into 'manure' to be used in crop cultivation. On a more positive note, recycling is relatively widely practised in several developing countries – though this is more likely to be due to economic necessity than to a conscious concern for the environment. One danger, though, is the use of some empty containers (such as pesticide containers, etc.) which may have traces of toxic chemicals, by local residents for carrying and preserving potable water, among other things. Also, absurd as it may sound, it has been recorded that certain workers are known to use the plastic linings from pesticide bags as raincoats during the rainy seasons (Colopy, 1995).

In any case, large volumes of very toxic materials are less likely to occur in most developing countries due to the scale of industrial production in such regions. However, even small quantities of carcinogenic or highly toxic chemicals released into the environment can cause significant health and environmental impairments. Consequently, there is an inevitable threat to human health and the environment due to the possible presence of highly toxic materials that are not managed in an environmentally sound manner.

In practice, governments of most developing countries may be operating on a philosophy of industrializing first, and correcting for the pollution problems afterwards. Indeed, because of the limited financial and technical resources available, little pollution control measures appear to have been implemented in most developing or newly industrializing countries. Only a very limited number of well-engineered disposal methods and waste management schemes seem to be available in such regions. Mostly the 'least expensive' methods of disposition of wastes seem to be the methods of choice for many of these regions. Such methods include open dumps (waste piles); shallow sub-surface burials; surface impoundments (for both liquid wastes and sludges); open burning; discharge to sewers and drains; and discharge to open land (usually farmlands). In all such methods, the wastes are exposed to long periods of severe weather conditions (including precipitation, evaporation, infiltration, erosion, photolytic reaction by exposure to direct sunlight, etc.), causing extensive migration of chemicals to soils, water, air and foods. Notwithstanding the fact that some processes such as photochemical reactions may be an effective natural process that results in rapid degradation and detoxification of some chemicals, in several other cases (e.g. relating to the pesticide dieldrin), a photochemical product of greater toxicity than the parent compound may result.

In general, waste management practices in most developing countries can at best be described as crude in most cases. This situation has been created as a result of inadequate or non-existent environmental policies and regulations, the lack of political will to prioritize environmental issues as part of the national political agenda, and the lack of technical environmental standards for pollution control and waste management. Furthermore, there has been an increase in the total amounts of waste generated from industrial activities – owing in part to an increase in the number of foreign industries that have been locating facilities in the regions, and in part to some of the activities associated with a so-called structural adjustment programme and diversification economic initiatives in certain parts of the developing and newly industrializing worlds. All these factors, coupled with the lack of technologically sound TSDFs in many of the developing countries, have contributed to the inadequate waste management situation in these regions.

3.2.3 CRITICAL ENVIRONMENTAL ISSUES IN DEVELOPING COUNTRIES

Systematic information on general waste management issues in most developing countries is scarce. Apparently the United Nations Environment Programme (UNEP) has been focusing on industrial environments (personal communication with Fritz Balkau, Senior Programme Officer, Industry and Environment Office, UNEP, December 1990 and March 1991 – among others), and therefore no information could be provided by the UNEP Office for Hazardous Wastes from the limited industrial base in several developing countries. But it should be recognized that although the quantities of hazardous wastes produced by most developing countries may be small, the potential problems associated with such wastes are **not insignificant**. In fact, pollution is a real problem in the developing countries today – potentially affecting human health, environmental integrity, and long-term economic development in many areas.

As a matter of fact, several environmental problems – including those annotated below – apparently engulf many developing countries; some of these are reported past issues, whereas others are perceived as probable problems due to the trend of environmental protection programmes that exist (or do **not** exist) in most of these countries.

Susceptibility to widespread environmental pollution problems. Several developing countries are susceptible to widespread pollution problems in view of the apparent inadequacies in their waste management and environmental protection programmes. For instance, it has been reported for Nigeria that industrial wastes, sewage and other waste materials have apparently been indiscriminately disposed of on land or into surface waters; additionally, oil spills have contaminated both surface water and groundwater. Furthermore, the intensive use of fertilizer by various government-sponsored agricultural establishments has led to heavy nitrate loading of waters in the Otamin watershed in south-eastern Nigeria (Egboka *et al.*, 1989). Amidst all these there is no enforceable legislation yet in place to manage the situation effectively. And Nigeria does not appear to stand alone in this type of situation, since this seems to be the *modus operandi* in several developing countries.

In fact, general waste management processes in the rural areas in particular are probably the most inefficient. For instance, in many rural communities in several developing countries, shallow pit latrines are used for disposal of human excreta, expired drugs and other chemical products – all of which can result in the long-term contamination of various media, including potable water supply sources. Additionally, pit latrines and open-space commercial toilets are crucial sources of fecal/pathogenic pollution.

Lately, additional problems with respect to environmental degradation are due to the dumping of toxic wastes from some industrialized countries onto certain developing nations. This is further to the improper control and management of imported hazardous materials and other virgin chemicals meant for local consumption/usage.

Inadequacies in the design of waste management systems. Apparently there are widespread inadequacies in waste management practices throughout numerous developing countries – judging from the abundance of open dumps, open burning, littered city streets, clogged stormwater sewer systems, etc. in some places in these regions. Controlled sanitary landfill has rarely been practised (Holmes, 1984), and the very high rainfalls in some of these regions tend to cause drainage and leachate problems that require special care. In most cases the usual method of disposal has been crude controlled tipping or semi-controlled tipping; the waste is generally used to reclaim and level depressed areas like ravines, old quarries and valleys (ibid.).

In general many waste sites are at best poorly designed, and uncontrolled leachate may be contaminating groundwater at several locations in these regions. A typical waste site operated in many developing countries is not professionally designed and has little to no information on the waste types, operating conditions, etc. In addition several of the sites may be located in poor hydrogeologic/environmental settings – including being located in rather permeable soils, close to the groundwater table and within proximity of surface water bodies and residential settings. Although disposal sites in the different countries within the developing world may each be unique, most of them probably have in common the potential for contaminating the environment by releasing toxic chemicals.

Water resource degradational problems from mining activities. Groundwater and surface water pollution may be a major problem in developing countries, particularly in areas of mining and extensive industrial activities. For instance, acid mine water containing trace metals is believed to be one important source of contamination, as manifested by the acid mine drainage problems at the Enugu coal mines of Nigeria (Egboka *et al.*, 1989). Egboka *et al.* (1989) has also recorded that many people from the Abakaliki mining district of Anambra State in Nigeria suffer from lead poisoning as a result of contamination of their water supply sources. This kind of situation is probably a common occurrence in many areas in the developing world. In fact, environmental media contamination from mining activities exists in a number of developing countries, and it is not unusual to find that waste streams from mines contain several hazardous components – including various heavy metals (e.g. arsenic, cadmium, lead, zinc) and their compounds.

Groundwater vulnerability from leaking underground storage tanks. Possible petroleum leakages from underground storage tanks (USTs) and oil pipelines may constitute an increasing threat to groundwater quality in many developing countries. For instance, leaking USTs are potential sources of contamination of soils and water supplies with benzene, toluene, ethylbenzene, xylene (BTEX), inorganic lead (from leaded petrol) and indeed other chemicals. Historically it is believed that there has been reported cases of 'loss of stock' from USTs used at fuelling stations in many developing countries, but no mitigative or remedial measures were known to have been implemented.

Consequences of multimedia pollution and cross-media contamination problems. It is apparent that significant amounts of hazardous wastes end up in municipal dumps. In some cases of close proximity to the tropical rain-belt, some developing countries must cope with frequent heavy rains. Thus, landfilled wastes are subject to the strong possibility of rapid leaching or even direct overflow into surface waters.

During the rainy season, wastes are washed into river basins and watersheds where they potentially contaminate soils, sediments, groundwater, surface waters and indeed the food chain. Contamination of water supply sources and even direct exposure of nearby residents could result, because in several cases, landfilling of hazardous wastes occurs close to industrial estates which are surrounded by poor neighbourhoods. Runoff from such sites can contaminate extensive areas. Polluted surface water can result in higher fish mortality. Where fishing on subsistence level provides a means of livelihood, significant drops in fish productivity due to pollution can have severe economic consequences on the community. Potential health risks from contaminated fish are also an issue in this case, since local residents may be consuming seafood with bio-accumulated levels of toxic chemicals that are present in the environment. Polluted irrigation water and agricultural lands pose both health risks and reduced crop yield; this potentially results in both economic losses and health impacts. Polluted potable water supply sources present an even more direct impact.

Lack of relevant environmental database and logistic systems. At the present time, there is insufficient knowledge of the hazardous waste situation in most developing countries. The importation, storage, use and disposal of hazardous materials or wastes are usually not adequately monitored. Furthermore, few developing countries have appropriate legislation covering the transport, storage or disposal of hazardous materials. But one of the biggest problems is the lack of environmental data, as well as the lack of an effective framework for assessing environmental impacts in the developing countries. In fact, some developing countries are experiencing rapid industrialization and a corresponding increase in hazardous waste generation. However, most of these countries have no legislation and/or enforcement policies concerning the management of such wastes, or any type of wastes for that matter. Also there is little to no data on the quantities of wastes generated or disposed of by most newly industrializing countries. The number of disposal sites are not documented either. Poor public health statistics also mean that in trying to develop an adequate information base regarding potential impacts of hazardous waste management in such countries, it will be difficult to relate particular episodes of a disease/illness to chemical exposures resulting from inadequacies in waste management processes. Thus it is more difficult to carry out any epidemiological studies associated with chronic chemical exposures in most developing countries because of the logistic and statistical problems involved – except for cases of accidents involving acute exposures. Without adequate information on all these, it becomes difficult to implement adequate corrective actions.

Many developing and/or industrializing countries share the environmental problems identified here. Some of the problems are common and transnational and indeed require regional or even international cooperation. The possible existence of shared and similar problems calls on the need for collective stewardship in resolving the environmental problems of regional issues.

3.2.3.1 An environmental dilemma?

Potential health problems due to waste mismanagement practices in many developing countries may be more critical than believed – especially when one considers the fact that such substances as asbestos and materials containing PCB may have been disposed of without due care and control. Indeed, human exposures to toxic chemicals may be occurring at higher levels than in the industrialized countries – representing a case of 'latent or silent killers'. Thus, unless efficient management methods are implemented to curb these growing problems, populations and ecological systems in such regions will become infested with ever-increasing risks.

It is noteworthy that the potential health risks, environmental impacts, and other long-term implications arising from inadequacies in waste management practices may exceed short-term savings (from avoidance of waste management expenditure) by several orders of magnitude. Apart from immediate and direct health and environmental threats, inadequate waste management practices could lead to the long-term contamination of the ambient air, soils, sediments and surface waters, groundwater resources, and biota. Furthermore, the cumulative effect of improper waste management practices will be a severe negative impact on any so-called structural adjustment initiative or economic development programme. These could all potentially reduce the overall future development capacity for a given region. Consequently, there is the need for a careful evaluation of any national policy and development programme in the developing countries, such that effective waste management practices can be incorporated in the overall agenda.

Several environmental dilemmas are indeed apparent to virtually all the developing nations and countries with economies in transition; a number of specific factors have been identified as potentially contributing to the lack of significant progress in resolving the environmental dilemmas in several of these countries, including the following (Asante-Duah and Sam, 1995):

- Lack of adequate legislative capacity to develop environmental statutes and promulgate environmental laws.
- Absence of comprehensive regulatory systems.
- Diminishing authority and capacity of some local/municipal governments involved in environmental policy and compliance monitoring.
- Lack of significant attention to environmental management issues as part of the overall national economic and political agenda.
- Lack of adequate financial assistance from international institutions and donor countries of the industrialized world.
- Apparent low priority that is given to the regions' deteriorating environmental situations by the international community (as reflected by the level of real assistance that is made available).
- Absence of effectual environmental education and training programmes, and lack of public awareness.
- Lack of public involvement in decision-making and econo-environmental programmes, and inadequate integration of environmental and socio-economic issues in policy-making.

Special attention should indeed be given to the factors identified here, as well as several other situation-specific issues. Such a focus will help address the environmental dilemmas of most developing or newly industrializing countries within the context of resolving global environmental issues facing the international community at large. In fact, whereas each of the countries in the developing world sees economic growth as a means of improving the quality of life and general welfare of its population, continued disregard or inadequate protection of the environment could jeopardize the long-term prospects for the very growth initiated.

3.2.4 THE NEED FOR ENVIRONMENTAL LAWS AND POLICIES

Even though the industrial sector in most developing countries is rather small by world standards, it is growing steadily – and most of the industries in these countries do indeed generate substantial amounts of hazardous wastes. In general the transformation of raw materials into final utilizable products involves the creation of some forms of wastes, and since most developing countries have some form of industrial base, these countries are not immune to hazardous waste problems. But most governments and policy-makers in these countries have in the past opted for industrial growth with minimum or no environmental controls. Consequently, hazardous wastes from industrial, agro-chemical and other sources are, invariably, improperly disposed of, leading to contamination of water resources and other media, and also posing direct health risks to populations in these regions. There usually are few, if any, programmes in place to help manage waste materials generated by industrial and other activities in most of these countries. Thus environmental degradation has already been occuring in many parts of the developing world. Such degradation will continue with consequential damage to the quality of life in these countries, unless some proactive measures are taken. Indeed environmental degradation could become a threat to the overall sustainable development in these regions.

There is a real need for developing and newly industrializing countries to set up agencies and regulatory authorities with the responsibility for the regulation and management (including the proper disposal) of toxic chemicals and wastes generated or existing in these countries. The applicable laws should, at the minimum, cover the following regulatory aspects and control needs for the particular country.

Toxic chemical inventory. Toxic chemical regulation and management should involve the collection of basic information on all chemicals in use in the country and the maintaining of an inventory of these chemicals; the identification and control of potentially harmful chemicals and materials; and the establishment of a notification process for all new chemicals introduced into the country.

Risk-benefit assessment. For all chemical uses in any one country, if the potential risk-benefit ratio associated with a chemical substance is greater than unity (i.e. if the risks outweigh the benefits), then a decision should be made to eliminate such material from the society, or at least to put stricter regulatory controls on its use in the country.

Legislative-regulatory protocols. Enactment, implementation and enforcement of regulations covering hazardous waste disposal and management should comprise a comprehensive system to monitor and control hazardous waste generation, transport, storage and disposal; the system should be designed to function in a manner similar to the 'cradle-to-grave' system used in the US and other industrialized countries.

Since the existence of laws and regulations without stringent enforcement mechanisms and deterrents will not ensure the proper management of hazardous chemicals and wastes in any of the developing or newly industrializing countries, there should be strict and deterring penalties established for violations of any implemented laws.

3.2.5 GENERAL ENVIRONMENTAL CONTROL NEEDS FOR DEVELOPING COUNTRIES

Indiscriminate disposal of hazardous wastes along with refuse in open dumps still seems to be the most common method used in a number of developing countries. Recognizing this, it becomes apparent that the serious incidents which forced the developed countries to regulate hazardous waste dumping activities could be repeated in the developing and newly industrializing countries with increasing frequency and severity in the future – that is, if adequate and effectual control programmes are not implemented promptly. In fact, the likely existence of significant waste streams of potential concern in many developing and newly industrializing countries suggest an urgency in addressing the issues associated with hazardous waste management in these regions. To this end, several control measures may be required to help abate the critical environmental issues that exist in the regions. These include the following.

Creation of environmental database systems. Apparently, no significant information exists on hazardous waste generation and management practices for most of the developing and newly industrializing countries; only limited national estimations are believed to exist for a few countries. It is important that information is gathered on waste management issues in all countries, to create systematic and consistent database systems that can be updated on a regular basis. This may be considered as a prerequisite for instituting effective environmental protection programmes, or should at least be seen as an important part of such programmes.

Recall that in 1972 the United Nations Conference on the Human Environment, held in Stockholm, recommended the setting up of a centralized register of data on chemicals likely to enter and damage the environment. Two years later the Governing Council of the UNEP decided to establish both a chemicals register and a global network for the exchange of information contained in the register. In 1976 a central unit for the register, entitled the International Register of Potentially Toxic Chemicals (IRPTC), was created in Geneva, Switzerland. Chemicals examined by the IRPTC have been chosen from national and international priority lists. Selection criteria include quantity of production and use, toxicity to humans and the ecosystem, persistence in the environment and the rate of accumulation in living organisms. IRPTC stores information that would permit an assessment of the risks and hazards posed by a chemical substance to human health and the environment. To take advantage of such existing systems, the developing and newly industrializing countries should institute Regional Correspondents for the IRPTC, so that chemicals of interest to their countries are compiled in an IRPTC Waste Management File, which can be updated periodically.

Implementation of a public educational programme on environmental issues. It is apparent that several communities in many developing countries are being impacted by a variety of potentially hazardous chemicals, and most of these countries have no plans, management programmes and instruments in place to help abate this situation. Also, although scavengers seem to contribute to recycling efforts in several developing countries, they generally lack knowledge about the potential dangers from their contacting hazardous materials during the sorting of wastes at disposal sites. Furthermore, naturally occuring toxins are potential threats to some regions in the developing world. For instance, aflatoxin (which is a mixture of very toxic chemicals potentially causing liver cancer) has been found in corn, rice, peanuts and other foods; the source of this toxin is a fungus, *Aspergillus flavus,* which occurs worldwide, but is especially common in warm countries, where crops such as corn, rice and peanuts are allowed to become wet after harvest. Similarly toxicity of many species of mushrooms consumed in some of these countries has been known for several years. But, until there is an outbreak of a disease that is directly pollution-related, waste management or environmental control is not usually seen as a priority in national policy due in part to lack of public awareness, and even due to some degree of ignorance of some policy-makers. In fact, ignorance of the potential harms of toxic materials still exists among much of the developing world – and public education may be a critical requirement to reverse the trend of affairs.

Design of appropriate waste management systems. In most developing countries, open waste dumps are apparently commonplace. With inadequate controls on what goes into such dumps, hazardous wastes are potentially an integral part of their composition. Some of these dumps may already be leaching chemicals and causing several water pollution problems, as well as affecting the health of scavengers and nearby residents and/or numerous other workers. There is the need to address this problem of controlling waste disposal methods (including waste segregation when necessary, and the use of engineered landfills). A site with good natural conditions is critical for even well-designed hazardous waste facilities.

Controlled landfill can be an economical and effective means of waste disposal. Provided careful site selection and operation procedures are followed, this can have minimum health and environmental impacts. However, follow-up actions are generally needed for urban waste sites, in order to monitor the continuing generation of leachates. In fact, because of the high rainfall and potentially sensitive aquifers in some of these countries, the control of landfill leachates and their migration through the use of well-engineered landfills is crucial.

Hazardous waste land treatment (HWLT) facilities may also be designed where extensive studies would have proven that, under prevailing climatic and other conditions, some specific wastes will be degraded, transformed or immobilized within the treatment zone above groundwater table; several engineering, environmental and economic factors should be considered for such a selection.

Incineration offers a more sophisticated process requiring a high degree of technology and high capital costs (especially if intended for toxic wastes where gas scrubbing equipment is usually required). Some wastes such as oils and organic solvents are preferably managed by incineration.

In general, adequate site selection should be carried out for the siting of well-engineered landfills and HWLT facilities for local uses, while a consideration is given to regional facilities for waste incineration purposes.

Development of responsible environmental policies. Little or nothing is known about what chemicals actually exist in the environmental media in many developing and newly industrializing countries. Neither is there any comprehensive policy for the control and regulation of hazardous materials, or for the disposal of hazardous wastes.

As part of the policy strategy, a survey on the environmental fate and toxicological information should be conducted for industrial chemicals (the results can then serve as a potential guide to policy-makers in developing countries when reviewing chemical usage in their respective societies). To illustrate the dimension of chemical misuse, it is noted that chemicals such as DDT (a bio-accumulative cancer-causing chemical and reproductive toxin) are believed to have been used as a crude 'aid' by some local fishermen in fishing from village streams in some places in certain developing countries. Also, it has been noted that in some other similar situations, indigenous people have caught fish by pouring an insecticide containing lindane (a highly poisonous chemical with serious health risks) into surface waters, and even when the contamination became apparent, absurd as it might sound they believed simply cutting off a fish's head made it safe to eat (Colopy, 1995). Appropriate evaluations should help bring about a renewed focus on such issues so that governments of these countries can re-examine general exposure of their populations to toxic chemicals. Ultimately, and ideally, it is necessary to control all chemical and infectious wastes and/or materials entering the environment, so as to prevent or limit potential adverse health effects and environmental impacts.

Institution of equitable international environmental standards for the modus operandi. The double standards displayed by some multinational corporations in their operations and marketing of consumer products, pesticides, etc., potentially contribute to the toll of health and environmental problems in some developing and newly industrializing countries (see, e.g. Colopy, 1995). There is the need for harmonized and non-discriminatory control measures relating to the export of hazardous materials to other nations, that are no less stringent than those applicable to the exporting country or international standards, whichever is deemed stricter. In this regard, several variables should be considered for developing regulations (including risk, economic, technical and cost-benefit factors) for the consumer products finding widespread use in several developing and newly industrializing countries.

In fact, notwithstanding their experiences elsewhere and their relatively better financial status, some multinational corporations and mining companies located in many developing countries may not be doing enough in environmental control programmes associated with their activities. This situation needs to change, however, if a sustainable development programme is to be born in many developing and newly industrializing countries.

Recognition of the values of environmental protection. The lack of financial resources and skilled manpower for hazardous waste management is a major concern. Hard currency restrictions make it difficult for most policy-makers to include environmental issues on the priority agenda; consequently, in most developing countries there is no legislation in place to guide the safe disposal of hazardous wastes. But, improperly managed disposal practices may reduce future development potentials of a country. Thus, although proper control of hazardous wastes may be seen as expensive today for developing and newly industrializing countries, it still may be cheaper and indeed safer to maintain a sustainable environment in order to avoid potential long-term costs and impacts.

In general, preventive programmes often cost significantly less than do curative programmes that might be needed in the future. Quantifying this difference in costs of prevention and cure could strengthen arguments for prevention. In fact, there is a general need to implement both curative and preventative environmental action programmes for the developing and newly industrializing countries. Control strategies should incorporate details on nature and amounts of chemicals and wastes, technical standards, and guidance for implementing waste management strategies.

Figure 3.2 shows a protocol that could be used to guide policy-makers in many developing and newly industrializing countries in the development of environmental control programmes that will effectively address the critical environmental issues originating from hazardous wastes present in these countries. In any case, the first priority will be to introduce legislation that establishes, in each country, a responsible agency for dealing with hazardous wastes, and that is further charged with the development of appropriate regulations. In this regard, the design of transport legislation (as an example) should require that only special vehicles be used for moving hazardous wastes and that only authorized carriers are allowed to transport them. There must be a strict system (such as a consignment notification procedure) to control the movement and disposal of hazardous wastes and prohibit the disposal to any site other than to designated hazardous waste disposal facilities. Also very important, national masterplans for hazardous

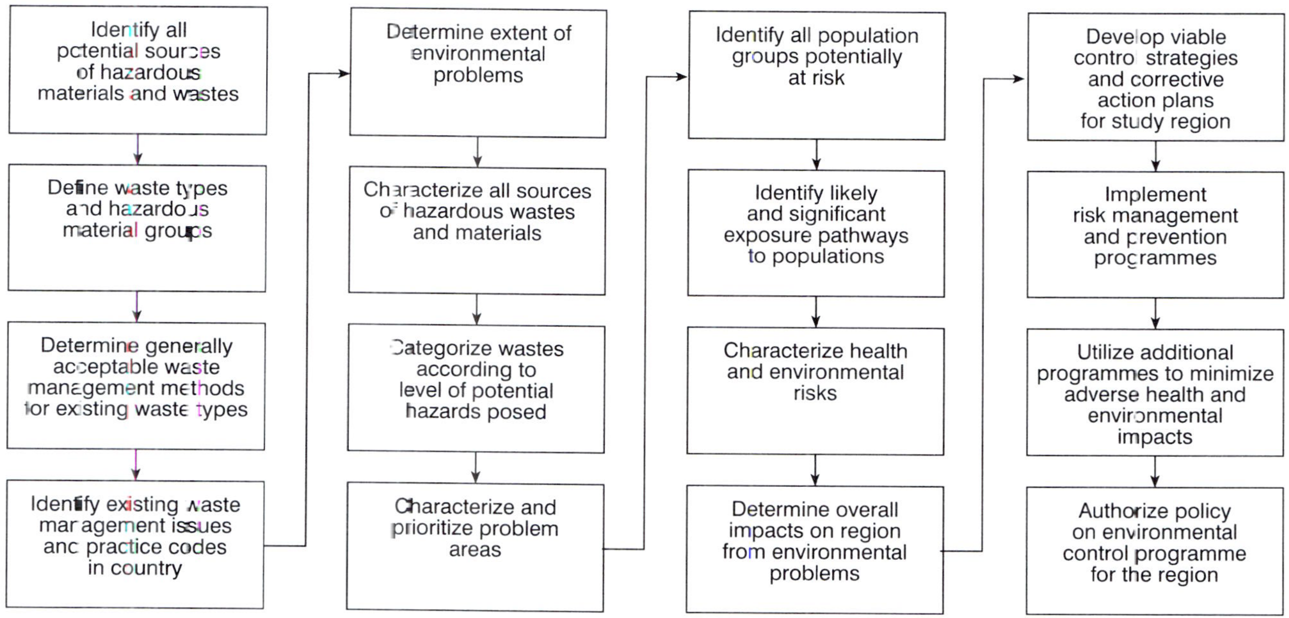

Fig. 3.2 Protocol for developing environmental control programmes for hazardous waste management in developing countries.

wastes (including accurate data on the transport and use of hazardous wastes, and the sources of wastes that are essential to the planning process) should be developed. Ultimately the proposed protocol can be adopted to facilitate the development of a comprehensive plan for the case region. The suggested protocol is based on the assumption that adequate financial resources will be made available for the implementation of the applicable or relevant environmental initiatives.

3.3 WASTE FACILITY SITING PROBLEMS AND RESOLUTIONS

Several aspects of hazardous waste management problems usually become apparent during the siting of hazardous waste TSDFs. Facility siting typically brings government, the private sector and the public together in an, often, emotionally charged atmosphere, where tough decisions have to be made. The consequence of not finding solutions or compromise to facility siting problems is that hazardous wastes could be handled illegally and/or inappropriately, which will invariably result in higher and longer-term costs for society. In fact, hazardous waste management facilities, being the kidneys of an industrial society, require a responsible system for dealing with them. Thus, regardless of the differences in their agenda, the similarity of basic interest (with regard to minimizing public health problems and potential ecological impacts) should make all stakeholders allies in the matter of hazardous waste management, rather than adversaries.

In general the most common attitude among the public is a very negative syndrome and resistance to hazardous waste management facilities being located in their community, especially because the presence of such TSDFs tends to lower property values in the area. There is also the psychological effects of real or perceived risks. Consequently, governments and industry must both understand and appreciate how and why individual homeowners feel threatened by the possibility of a TSDF in their locale. One possible way to resolve these types of problems is by public education, dialogue, mediation and striving to win public trust in facility siting. In fact, the resolution of any conflicts is not only desirable – it is possible. There is a necessity to involve the public in decision-making and notify people about what has already been done, and develop an understanding of mutual needs and overlapping interests.

To facilitate a community acceptance of a TSDF in their neighbourhood, the community needs to be assured of an increase in the number and security of jobs, the preservation of a satisfactory quality of life, the maintaining of property values and the reservation of the right to keep undesirable facilities out of their neighbourhood. Ultimately national governments assume the responsibility of assisting their industries in dealing with the wastes produced or imported, as well as safeguarding the health and well-being of their citizens. Typically several determinations are important for evaluating and addressing the potentially significant environmental and waste management issues, and due consideration of the following items can help accomplish this overarching mandate of ensuring public safety.

- Establishment of a complete knowledge of waste characteristics – including waste types, degree of hazards, waste compatibilities, and the ability to segregate ignitable, reactive or incompatible wastes. Wastes should be well characterized so as to separate incompatible wastes and to select suitable treatment and disposal techniques.
- Determination of the fate and transport characteristics of chemical constituents of wastes and their projected degradation products (including an identification of anticipated reaction or decomposition by-products).
- Identification of the critical media of concern (such as air, surface water, groundwater, soils and sediments, and terrestrial and aquatic biota) that could be impacted under different release scenarios.
- Evaluation of potential exposure pathways of waste constituents, and the potential for human and ecosystem exposures.
- Assessment of the environmental and human health effects of exposure to the hazardous wastes, if such wastes should reach critical human and ecological receptors.
- Characterization of disposal sites – to include site geology, topography, hydrogeology and meteorological conditions.
- Implementation of substantially stringent standards regarding the information and statistics to be provided by industry on its chemical products and emissions/discharges.
- Analysis of the leaching properties of solidified hazardous wastes and implementation of higher standards for waste management and landfill design.
- Design and application of a coherent regulatory framework to cater for hazardous waste management programmes – to include a review process for permits issued to industrial plants under environmental legislation; definition of general strategies and procedures for handling chemical products; and the adoption and use of an effectual waste tax system.
- Waste sorting at source, and a separate reuse and recycling programme that allows leftover wastes to be disposed of without endangering the environment.
- Stricter application of the substitution principle as regards chemical products and intensification of work on 'eco-labelling' (such as the green swan scheme in Europe) and environmental product information, to make it easier for consumers to make decisions on choosing or buying more environment-friendly products.
- Elaboration of a code of ethics to guide scientists, governments and industry in the proper handling of generated wastes and related environmental pollution problems.

- Adoption and use of risk- cost-benefit analysis as part of the decision-making protocol in the management of hazardous waste TSDFs.
- Determination of extent of service area for any proposed waste facility – that is, whether it will handle wastes from local industry only or be a regional facility involving possible transboundary movements of wastes in a waste trade agenda. This should include what the overall economic and environmental impacts on a country and/or region are projected to be.
- Confirmation that the location proposed for a waste facility is suitable, given environmental, social and economic concerns – including non-proximity to populations, ecological systems, water resources, etc. For instance, active seismic areas and floodplains should be excluded from the siting of waste management facilities.
- Indication that any proposed waste facility represents the best available technology (BAT) or BATNEEC for handling the particular wastes.
- Incorporation of contingency plans and emergency procedures in the design of the waste management plans.
- Inclusion of effective long-term monitoring and surveillance programmes – including post-closure maintenance of facilities as a necessary part of the overall TSDF design plan.

It is noteworthy that special attention should also be paid to the interest and survival of small-scale industries within the framework of implementing an environmentally sound hazardous waste management programme. In any case, once the determinations and diagnostic assessments of potential environmental problems are completed for any given region, plans can be made towards effective corrective actions where warranted. The overall goal will be towards the minimization of health and environmental impacts, and the concurrent optimization of economic, social and psychological impacts on the region.

It must be acknowledged here, that a formal environmental impact statement is not presently required in most countries in the case of siting hazardous waste TSDFs. At the same time, in the countries where the relevant regulation is already in existence, there is very often a mismatch between the type of information generated by the official organizations proposing the hazardous waste project and the type of information which citizens need to evaluate these facilities. The result is very often the 'not-in-my-backyard' attitude (or so-called NIMBYism), and some groups of citizens have sought court injunctions against such facilities. Failure to provide the necessary information to the public, as may be required by the normal permitting process, tends to breed distrust and fails to elicit the cooperation of citizens and local officials who will feel short-changed and imposed upon. In order to fill this information gap, it is advisable that proponents of hazardous waste disposal facilities provide correct environmental impact information upfront that addresses potential health and safety concerns which are likely to be raised at public hearings. In fact, an informed public will generally be more of an asset than a liability in a hazardous waste siting project.

3.4 PRESCRIPTIONS FOR GENERAL WASTE MANAGEMENT AND RELATED ENVIRONMENTAL MANAGEMENT PROGRAMMES

In one classic example of a waste mismanagement situation, when a committee of the British parliament reported on waste disposal in 1981, reference was made to the fact that, seven years earlier, an operator at a landfill site near London had died as a result of breathing toxic fumes resulting from the accidental disposal of two incompatible wastes at the same location. Another classic example a world away involved disposal of chemical wastes for about three decades, at Love Canal, a disposal site in New York, USA. Subsequent use of the site apparently resulted in residents in a township in the area suffering very severe health impairments and, it has been argued, in several children in the neighbourhood being born with serious birth defects (see e.g. Gibbs, 1982; Levine, 1982). In general all waste disposal operations do constitute a potential source of risk – either immediate or future. Effective risk management strategies should therefore become an integral part of all waste management programmes.

Ideally, environmental management programmes for both industrialized and developing countries should incorporate an assessment of risks faced by these countries from all chemical material sources. A comprehensive agenda for dealing with the environmental problems in most sovereign nations should, among other things, accommodate the following prescriptions specified below.

Establish a national council on hazardous materials/wastes control. A comprehensive national waste management strategy for hazardous waste management issues should be developed for all countries. This should include the development of a good information base on waste quantities, and the prioritization of potentially contaminated areas. Programmes should be instituted for the identification of site-specific problems, and to map out potential problems of widespread uncontrolled chemical disposal practices from industrial, municipal and agricultural policies. A national council on hazardous materials/wastes control may be established that has a division responsible for monitoring waste management policies and practices. Such a council should comprise members from local, district, regional, federal, etc. outfits including representatives from health and environmental protection ministries. The national council should also coordinate and exchange information with counterparts in other countries in a region.

Monitor and control the cycle of overall chemical usage. All sovereign nations should introduce comprehensive and pragmatic laws to provide for the regulation of hazardous waste problems, and also the use of other toxic materials in general. There should be adequate regulations and management schemes for toxic chemicals in general that find widespread use in the respective countries. These should include collecting basic information on all chemicals from manufacturers, processors and importers – to help build and maintain an inventory of such chemicals. Information collected should include their toxicity and potential harmful effects to humans and the environment. All new chemicals should be permitted, notified and inventoried. In particular, policy-makers should be wary of chemical substances such as pesticides banned in other countries.

All chemical materials (not just hazardous wastes) should be subject to the monitoring protocols to prevent abuse; this will also help offset attempts by agents to become involved in waste trade activities under the pretext of possible sham recycling programmes.

Establish waste management database systems. A survey must be carried out to determine the status of hazardous waste management in sovereign countries. A database must be established that includes the chemicals produced by the various sectors of the economy. Information should be gathered for all establishments in these countries, in order to create a systematic and consistent database that can be updated on a regular basis. Such a database may be tied in to the IRPTC Waste Management File maintained by the UN.

Create a national 'waste site atlas' to aid policy decisions. There is the need to identify and diagnose hazardous waste disposal sites, investigate potential high-risk sites, implement clean-up programmes for priority sites, and furthermore to develop proper waste management strategies and disposal sites for future waste streams. Thus general environmental impact assessments should be conducted to determine where pollution or dumping has occurred. In fact, all sovereign nations should carry out a thorough and country-wide inventory of all waste dumps. The inventory should consider both legal and possible illegal waste dumps, industrial (past and present) locations, etc., and indeed all suspect locations in a country. This can lead to the creation of a national 'waste site atlas' in which sites can then be prioritized for remediation and risk management.

Institute adequate regulatory controls by implementing a 'cradle-to-grave' waste management system. There is the need to design a comprehensive system to monitor and control hazardous waste generation, transport, storage and disposal. A 'cradle-to-grave' system similar to that used in the US and elsewhere could be adopted; this system includes the maintenance of an inventory of wastes throughout its life-cycle, establishment of appropriate standards for hazardous waste TSDFs, implementation of adequate monitoring schemes, etc.

In the context of a hazardous waste trade, no matter how strict a regulation to control or ban such trade, dumping may continue. This means that additional activities such as the so-called 'Dumpwatch' established by some African countries is very important and should be intensified. But first, this calls for the creation of regulatory agencies with strengthened capacity for enforcement and also monitoring capabilities; it should be ascertained that enacted environmental laws are enforced, and not widely ignored. Also modest rewards or award schemes should be instituted for those who give tip-offs for the arrest and conviction of would-be waste dumpers and culprits violating environmental laws.

Encourage environmental and waste management research. Similar to what most industrialized countries have in place, developing and newly industrializing countries must also encourage research into several aspects of environmental management – including the development of standards, protocols and practice codes for hazardous waste management; environmental monitoring; and the regulation of hazardous waste disposal – that are appropriate to their types of environmental setting, rather than depending on wholesale transfer of already established technologies and methodologies from the industrialized world. Considering, for example, that in many developing countries the high rainfall (due to intense tropical storms) increases the chances of contaminating the groundwater and even surface water resources from torrential rains, it is important to develop strict standards for landfills, especially with respect to site selection (to include extensive investigation of geological suitability), site development, site operation, pre-treatment of wastes to be dumped, waste types to be accepted, and post-closure management and monitoring.

Include waste management plans in development programmes. Project financiers (such as the World Bank) of programmes in developing and newly industrializing countries should make it a general requirement to include sound waste management programmes in industrial and related projects (in as much as environmental impact assessment (EIA) is becoming an integral part of most development projects). Such management strategies should receive continuing monitoring, technical assistance and training programmes.

Consider the efficacy of regional waste management facilities. All countries should investigate the feasibility of establishing general regional or subregional waste management facilities, to respond to economies of scale and to ensure the highest possible environmental standards. Since several TSDFs have large fixed costs (especially for handling high-hazard wastes such as PCBs that require high-temperature incineration), high volumes of wastes may have to be processed in order to break even.

Waste exchange systems within neighbouring countries should be established and encouraged, but this should be under strict monitoring programmes. In this regard, countries generating small quantities of hazardous wastes should view waste exports as the best economic solution to local waste management problems that will not compromise on environmental sustainability; such countries may invest in joint ventures to build regional hazardous waste facilities at appropriate locations in the region. The selection of appropriate locations may be based on several factors – including the degree of industrialization. The prevailing contributions of GDP from industry in participating countries can, among other factors, serve as a barometer for preliminary assessments in siting regional facilities for hazardous waste management; this criteria should, however, be used with caution since other important and dominating factors may have to be invoked prior to designating a locale or country to accept particular types of wastes, irrespective of the volumes of production or percentage contribution of wastes from such a country.

Develop effective public education programmes. Increased public awareness is important to building consistent and pragmatic environmental programmes. The general perception in several countries (especially in much of the developing world) is that environmental problems are not priority issues at present. But the reality is that most of these countries face potentially serious environmental problems which need to be addressed sooner rather than later. Public awareness is a strong determinant of the political will to adopt and enforce environmental policies. The awareness and understanding of the long-term implications of improper disposal of hazardous wastes remain limited. Furthermore, because related groundwater contamination and health risks are not well understood, the high costs of appropriate waste management and disposal plans are difficult to justify politically in the face of other priorities.

Non-governmental organizations (NGOs) and the mass media can play an important role in raising public awareness. These avenues should be used to generate the environmental awareness that may be needed to cultivate a political will to draw up environmental policies for sovereign countries. Also public education through schools, organizations and the media may have relatively low financial and political costs, and yet can be a very powerful means of increasing public awareness and thereby strengthening political support. A programme for public education, comprising environmental education from preschool through universities, is therefore imperative.

Adopt waste control measures via pollution taxes and subsidies. Market forces may be used to encourage environmental protection by rewarding con-

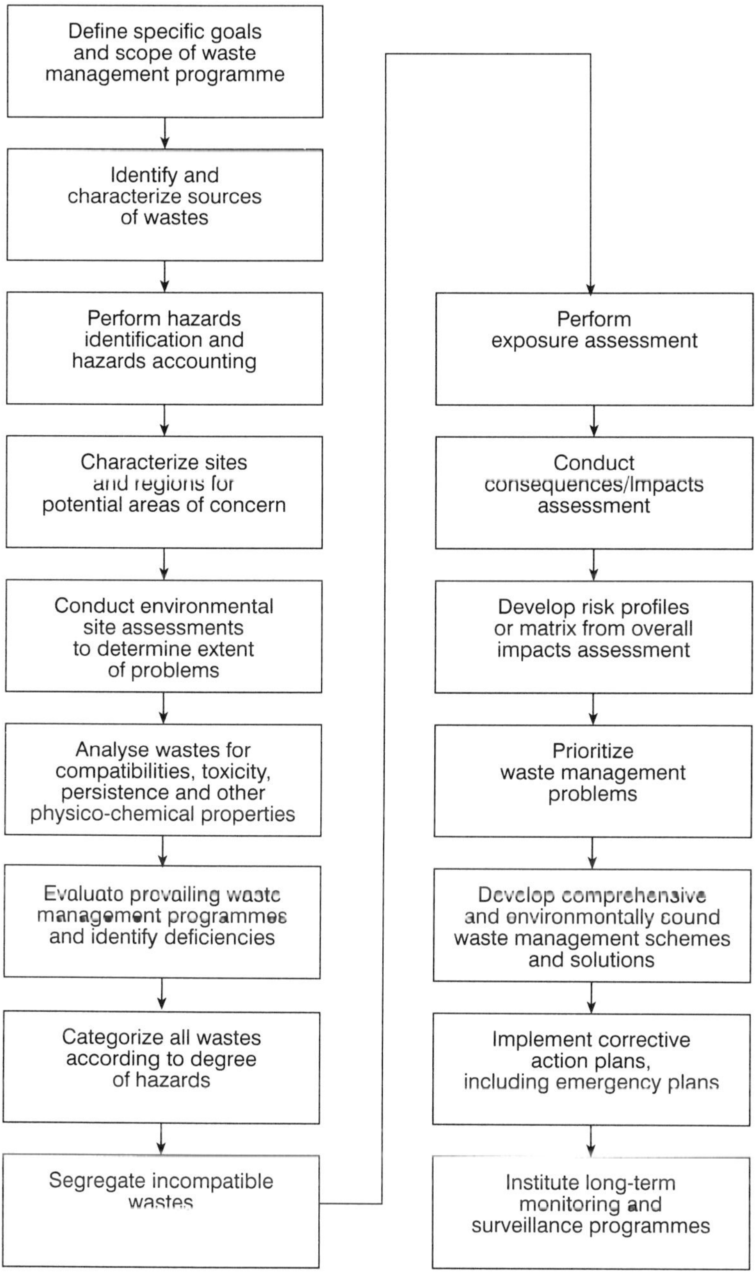

Fig. 3.3 Process flow for general waste management planning.

sumers and businesses that conserve, and penalizing polluters. As part of the relevant programme, special attention should be given to refuse collection, recycling and appropriate disposal in the waste-generator countries. Also incentive systems should be developed to encourage pollution abatement, waste minimization, recovery and recycling. Appropriate economic policies should be instituted to provide economic/financial incentives.

As a general guiding principle, sovereign nations should take adequate measures to encourage waste minimization, recycling and proper disposal of wastes. Materials and energy recovery and reuse should be encouraged. Tax/subsidy schemes that give incentives to firms that implement adequate pollution prevention strategies may be instituted. This means that despite stipulations of the prevailing regulations, if a firm goes below acceptable discharge limits, or does better in environmental control measures than stipulated by the regulations, then it could be rewarded by being allowed to take a 'tax holiday' or tax cut while all others and competitors pay pollution taxes.

These recommendations should generally aid policy-decisions on environmental control programmes in all sovereign nations. Notwithstanding possible financial constraints, it is pertinent that individual countries allocate some resources to develop comprehensive solutions compatible with the environmental problems in the specific countries; this may help avert potential long-term problems – especially where high-risk wastes may be involved.

Figure 3.3 shows a comprehensive flow diagram for designing a general waste management programme. The cost for such a scheme could be restrictive, especially in the developing and newly industrializing countries, and it may be that the only way to finance the programmes would be through funding from private enterprises. However, should this be necessary, strict governmental monitoring, control and enforcement should be maintained to minimize the chances for possible abuse.

The transboundary shipments of hazardous wastes

4

The production of large quantities of wastes globally has created a new commercial activity involving the transfrontier movements of hazardous wastes, intended to be managed at economically attractive TSDFs. In fact, huge quantities of hazardous wastes apparently travel the world in search of 'acceptable' TSDFs. For instance, millions of tonnes of potentially hazardous wastes cross OECD national frontiers each year on their way for recycling or to TSDFs because there is no local disposal capacity for these wastes, or because legal disposal or reuse in a foreign country may be more environmentally sound, or managing the wastes in the foreign country may be less expensive than at home.

Two key questions generally arise in the evaluation of the transfrontier movements of wastes: 'What types and quantities of wastes are likely to be prime candidates for export?' and 'What costs can waste generators expect to avoid, if the decision is made to export the candidate wastes?' It is apparent that wastes that cross frontiers destined for disposal in another country are likely to be wastes considered very hazardous, i.e. requiring incineration or physico-chemical treatment in the waste generator country. On the second issue, it is worth emphasizing the fact that costs for the legal disposal of hazardous wastes are high and still increasing, especially in the OECD countries. Capacity for legally disposing of hazardous wastes into or onto land is becoming more scarce. In fact, an OECD analysis of the prevailing situations concludes that the legal disposal will almost certainly become increasingly costly as a function of time (OECD, 1993). Also in situations where regulations require that certain types of wastes be destroyed by incineration, the siting of the relevant treatment facility is becoming more difficult. Furthermore, wastes considered candidates for export to other countries generally are restricted from legal dumping into the oceans. Thus, even within Europe, for example, journeys of up to 800 km can be contemplated by a waste generator in order to reduce total costs of disposal. But it must be acknowledged that the improper management of these wastes in countries with less than adequate technologies poses significant risks to human health and the environment, and potentially threatens development in such countries.

4.1 STIMULI FOR THE TRANSFRONTIER MOVEMENTS OF WASTES

It is estimated that hazardous wastes generated by US industry alone between 1970 and 1989 jumped from under 25 million tonnes a year to about 500 million tonnes a year (Figure 4.1); in fact, these figures may be much higher. Next to the US, Western Europe represents the greatest centre of hazardous waste generation in the world. In the OECD European area, the estimate for hazardous wastes ranges between 30 million and 45 million tonnes per year. It is estimated that the EC's annual production of hazardous wastes is about 20–30 million tonnes; another approximately 20 million tonnes are produced annually from the other OECD countries (Hilz, 1990). Central and Eastern Europe are estimated to generate about 6 million tonnes of hazardous wastes annually, and the developing countries probably contribute about 6 million tonnes annually to the global produc-

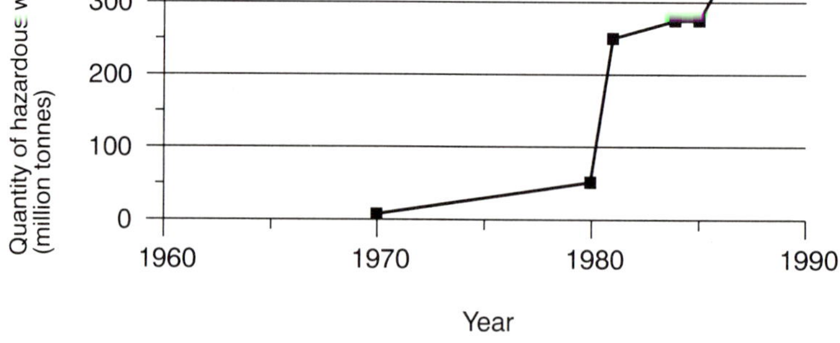

Fig. 4.1 Trend in the quantities of hazardous wastes generation in the United States between 1970 and 1989. (Source: Based on information from various sources, including the following – 'The Hazardous Waste Consultant', 7(5), 1989; Hilz, 1990; Asante-Duah, 1991.)

tion of hazardous wastes. A further 25–30 million tonnes of hazardous wastes are estimated to be produced each year in the former USSR region. Given the uncertainties associated with such basic statistics, it is difficult to come up with a good enough estimate of how much hazardous waste is indeed being shipped overseas.

There are a number of potential stimuli that cause waste generators to consider export as a preferred alternative for managing their wastes. Box 4.1 lists the critical factors contributing to the general transfrontier movements of wastes in search of alternative recycling, treatment, storage and/or disposal sites or facilities. Generally much of the wastes exported are believed to be highly toxic and dangerous; these include heavy metals (e.g. cadmium, lead, mercury), organic chemicals (such as PCBs and dioxins), inorganics such as cyanide, industrial solvents, waste pesticides, paint by-products, incinerator ash, sludge, spent chemicals and used storage/lead-acid batteries.

Box 4.1
Critical factors contributing to the transboundary shipment of wastes

- Overall stringent environmental laws in most industrialized countries.
- Tightening of specific laws, regulations and policies concerning disposal of certain types of wastes.
- Rising costs of hazardous waste disposal in the home country where wastes are generated.
- Diminishing domestic capacity for disposal of certain types of wastes.
- Potential future liability for any damages caused by wastes disposed of domestically.
- Market opportunities elsewhere for materials which can be recovered, reclaimed or recycled from wastes otherwise destined for 'final' disposal.
- General economic growth which may result in more total generation of wastes.
- Economies of scales associated with treatment, storage and disposal facilities (TSDFs).
- Existence of TSDFs which may serve several countries.
- Existence of an appropriate TSDF in a foreign country which is closer than a similar facility in the home country.
- Lack of environmentally sound TSDFs in some waste-producing countries.

Invariably the transfrontier movements of hazardous wastes across countries tend to reflect the degree of stringency of environmental regulations adopted by those countries for the management of hazardous wastes, with much of the flow having been directed towards nations with inadequate or non-existent legislative frameworks for dealing with environmental and/or hazardous waste issues. However, lately, political and diplomatic factors seem to be working to curb the uncontrolled and 'unidirectional' transfrontier movements of hazardous wastes. The US EPA's International Activities Office, among others, recognizes that waste export is not only an environmental issue, but also a foreign policy issue (CIR and Moyers, 1990). Furthermore, with most countries becoming more environmentally conscious, and with the rest of the world watching, the types of uninformed waste trade schemes that have taken place in the past are likely to be minimized, if not eliminated completely. Consequently, the type of scenario whereby 'hazardous waste from the industrialized nations frequently has a one-way ticket to the developing world' (CIR and Moyers, 1990) is likely to die off in the long-term. Nonetheless, since waste generators will normally seek the least cost available for the legal disposal of their wastes, it is expected that, if export avenues are available, legal and less costly than disposal in the home country, then waste export will become a likely choice.

4.2 THE 'WASTE TOURISM': WHERE ARE THE WASTES TRAVELLING?

Although the records of waste movements are incomplete and rarely consistent, an annotation of some typical situations are sufficient to understand the magnitude of such a flow. For example, it is believed that the US shipped about 100 000 tonnes of hazardous wastes abroad in 1987, and about 140 000 tonnes by 1989 (CIR and Moyers, 1990). Also a major route for waste movement has been from Western to Eastern Europe (albeit there are no accurate statistics on exactly how much waste is actually exported to all countries, since the problem does not seem to be well monitored).

Of particular interest in the 'waste tourism' schemes are the movements of hazardous wastes. In recent years, more than 4 million tonnes per annum of hazardous wastes have been moving across national frontiers in the OECD countries alone, as reflected in the early 1990s' data on the import and export of hazardous wastes within the 'traditional' OECD Member countries (Table 4.1); the data represents actual amounts in the most current year that was available to the authors at the time of writing. In fact, there is a rather well-developed waste trade connection between several European countries in particular. Within the western hemisphere, the movements of hazardous wastes are seen to parallel the patterns of regulations, because wastes move from countries which have adopted

Table 4.1 Early 1990s hazardous waste statistics for selected OECD countries. (Source: OECD, 1994.)

Country	Total annual production ('000 tonnes)	Annual transfrontier movements ('000 tonnes)	
		Imports	Exports
USA	180 000	50[a]	118.4
Canada	6 080	143.8	137.8
Germany	6 000	62.6	522.1
France	3 958	458.1	10.6
Italy	3 246	0.0	20
UK	2 940	54.1	0.9
Spain	1 708	82.3	20.2
Portugal	1 043	82.3	2.0
Netherlands	1 040	199	195.4
Switzerland	736	6.4	126.6
Japan	666	0.4	40
Austria	620	111.6	82.1
Sweden	500	47.2	42.6
Greece	450	–	0.3
Australia	316	9.0	91
Finland	314	9.9	19.2
Turkey	300	0.0	0.0
Norway	200	0.0	16.5
New Zealand	110	0.0	0.2
Denmark	106	16.4	9.2
Iceland	5	0.0	0.2
All OECD countries	237 388	2 262.3	1 870.8

[a] Vallette and Spalding, 1990; CIR and Moyers, 1990

and enforce more stringent standards or impose higher costs for the management of wastes, to countries with less stringent regulations or those which have lower waste management costs.

Globally speaking, there is no comprehensive inventory on hazardous waste volumes that may have crossed international borders in the past – this having been a less controlled and/or less monitored area over the years. Purposeful misclassification of wastes has further inhibited a good appreciation of the extent of the current trade practice; frequently the nature of the hazardous wastes is not fully revealed to the receiving country. It is believed, however, that more than 5 million tonnes of toxic wastes probably cross national borders each year, with a significant share of this volume being illegally dumped. Further evidence that the magnitude of these transfrontier movements may be far greater than recorded has become apparent in recent years, following the discovery of illegal flows of hazardous wastes from countries in the North to those in the South.

In fact, it has become apparent that foreign waste brokers have attempted to ship some form of waste to most developing or newly industrializing countries in the past, or contemplate doing so. The greatest concern in 'waste tourism' is in regard to waste exports in the direction of 'least resistance'. Typically the patterns of hazardous waste movements show two main directions of flow: from North to South and from West to East. For example, in 1988 more than 1 million tonnes of hazardous wastes were exported from Western Europe to East Germany alone; the major exporter countries included Austria, Belgium, France, Italy, the Netherlands, Switzerland, the UK and the then West Germany. Several developing countries have also become vulnerable to the waste export practices/schemes for various reasons. For instance, just by virtue of their proximity to the US, which is the world's largest waste-producing country, most Latin American and Caribbean countries became easy targets for the international waste brokers/traders in the US. But notwithstanding distance, Africa and Asia have been equally preferred targets. The irony of all these is reflected by the observation from Greenpeace, the international environmental pressure group, that, in the global waste market-place, the most attractive targets for waste dumpers are the countries that are furthest removed from the economic benefits of industrialization (Vallette and Spalding, 1990). In fact, prior to the international control measures (see Chapter 5), Europe alone had been exporting – legally – about 120 000 tonnes of hazardous wastes to developing countries each year.

Greenpeace has indeed been monitoring the transfrontier movements of hazardous wastes worldwide since 1986, and this has enabled it to compile an inventory of waste trade operations (Vallette, 1989; Vallette and Spalding, 1990). Up-to-date information on the status of these operations is published in the *Toxic Trade Update* newsletter and also made available on the computer network *Econet* (accessed from haz.net). It is apparent that some of the waste shipments taking place disregard national regulations and international conventions on the transboundary movement of hazardous wastes. A large part of the shipments also find their way around regulations by designating the wastes for recycling

4.2.1 PAST HAPPENINGS: THE FINE LINE BETWEEN ILLEGAL DUMPING AND LEGALLY NEGOTIATED BUSINESS TRANSACTIONS

According to documentation collected by Greenpeace, over 3 million tonnes of wastes were shipped from the industrialized world to the less-developed nations between 1986 and 1988 alone (Vallette, 1989; CIR and Moyers, 1990). Some of these shipments apparently had been made with the approval of governments of the receiving countries; for many other shipments, approval was never sought or obtained prior to export. Some of those countries which actually negotiated to receive the wastes had done so because the amounts of money to be earned from waste imports tend to be rather large; thus, despite the potential health and environmental risk implications, some nations have been tempted to enter into this kind of risky trade activity. Interestingly, a few industrialized countries have had formal waste import schemes in existence for some time now – except that activities in such countries have generally been formal and well orchestrated.

Examples of noted shipments or planned shipments of toxic wastes to selected countries are briefly discussed below. Whereas this is by no means complete, it is reasonably representative of the symptomatic scenarios prevailing in the hazardous waste market-place. It also reflects the potential savings for some waste brokers.

Albania. More than 500 tonnes of pesticides containing formulations of lindane and other organochlorine compounds and herbicides with high concentrations of dioxins were exported from Germany into Albania between 1991 and 1992. Disposing of such wastes in Germany would have cost about DM 8000 per tonne – several orders of magnitude greater than exporting it to Albania. Subsequently, however, the German Environmental Ministry agreed to take back the wastes; the cost of remediation has been estimated at about DM 10m. (Greenpeace, 1993).

Angola. A waste trade scheme was reported to be under negotiation between the Angolan government and a Swiss waste broker in 1988. This was supposedly to bring over 5 million tonnes of industrial wastes into Angola for incineration, storage and disposal in exchange for $2b., further to the provision of social infrastructure and jobs (Vallette and Spalding, 1990). Subsequently, this deal was retracted when Angola refused to put their signature to it. Incidentally, however, Angola is believed to have previously accepted other toxic waste shipments from Europe.

Benin. Benin has consistently entered into several transactions with respect to the waste trade. For instance, in a contract signed in January 1988, a European firm called Sesco Limited had been planning to dispose of hazardous (including nuclear) wastes from Europe and North America in Benin for a period of up to ten years. Sesco planned to ship between 1 and 5 million tonnes of wastes per year, for which it was to pay Benin approximately $2.50 per tonne of waste, giving the Benin government potential earnings of up to about $12.5m. a year as well as provide some additional investments locally (Vallette, 1989; Vir, 1989; Vallette and Spalding, 1990). Subsequently the government of Benin tried to cancel the contract, but were unable to return advance payments taken – although it is believed to have been cancelled eventually. Also, several tonnes of radioactive waste are believed to have been dumped in Benin by the Soviet Union and France. In fact, there have been several other allegations made against the government of Benin about entering or trying to enter into contractual agreements to import hazardous and radioactive wastes from several countries.

Brazil. Between 1987 and 1989, Brazil actively participated in the international waste trade. Much of the wastes imported into Brazil contained arsenic, cadmium, lead and other heavy metals. Several of the shipments of toxic wastes had come from the US and supposedly were meant for recycling and disposal; other proposed shipments from Europe in 1989 had been rejected (Vallette and Spalding, 1990).

Congo. It is recorded that Congo nearly became a major recipient of industrial wastes from the US and Europe in early 1988. The Congolese government, it is believed, was to be paid an average of about $31 per tonne of wastes delivered, with the entire contract worth some $176m.; the deal would have brought over 5.5 million tonnes of wastes into Congo over an eight-year period (Vallette and Spalding, 1990). The types of wastes that were to be shipped and dumped in Congo's Diosso Gorge included solvents, paint and pesticide sludges, and several other chemical wastes from the US and Europe. However, the agreement was later retracted by Congo and the deal rejected; some responsible Congolese officials were subsequently prosecuted for their role in arranging this deal.

Guinea. In February 1988 Guinea received about 15000 tonnes of toxic incinerator ash shipped from Philadelphia, US, that was believed to contain several toxic chemicals (such as aluminium, arsenic, chromium, lead, mercury and dioxin-based organic products). Indeed, this waste (originally branded as fertilizer) was first sent to Panama for use in constructing a roadbed, but had been rejected by Panama's Ministry of Health and sent back to Philadelphia. Guinea, however, had reportedly accepted it for disposal on the Guinean island of Kassa near the capital city, Conakry, at a compensation rate of about $40 per tonne (Third World Network, 1989; Vir, 1989; Vallette, 1989; CIR and Moyers, 1990; Vallette and Spalding, 1990), where it was eventually to be used for constructing concrete-like bricks. But this deal sparked international fury; following some diplomatic feuding, it was arranged for the ash to be removed and returned to the US. However, on the way back to the US, a number of unsuccessful attempts where made to dump on other African soils in Senegal, the Cape Verde Islands and Guinea Bissau (Vallette, 1989; CIR and Moyers, 1990).

Guinea-Bissau. Guinea-Bissau negotiated a $120 million a year (more than its total annual budget) contract to store industrial wastes from other countries, but public outcry forced the government to rescind its decision (Third World Network, 1989; Vir, 1989; CIR and Moyers, 1990). Again, because of public outcry and pressure, the Guinea-Bissau government, in May 1988, postponed a deal with the US and European waste brokers that would have brought some 15 million tonnes of foreign industrial wastes into the country over a five-year period (Vallette, 1989).

The proposed payment of about $40 per tonne of waste would have given the country a total potential payment of $600m. (which is four times the country's GNP, and twice the country's foreign debt). The wastes would have been dumped in landfills near Guinea-Bissau's border with Senegal. The whole deal was believed to have been masterminded by an Italian waste broker who has arranged numerous shipments of Italian wastes to developing countries.

Hungary. Waste brokers in Austria signed a contract in 1985 with a Hungarian company, Flexum, for the export of 60 000 tonnes of wastes (purported to be, and declared as domestic wastes) containing large amounts of different toxic materials. As a result of the deposition of these wastes onto Hungary, the drinking water supply system of a township suffered immense deterioration a few years later, and there was a need for reconstruction of the whole system. In yet another case, an exhausted oil consignment, containing high concentrations of chlorinated compounds has been imported from Germany by a Hungarian company, Flexiko, to be incinerated at a thermal plant. Such 'experimental' programmes have since been banned by the Hungarian authorities.

Liberia. Several illegal dumping activities, as well as disposals resulting from some pseudo-negotiated deals, are believed to have occurred in Liberia for some extended periods of time – especially of wastes originating from the UK (Vallette, 1989; Vallette and Spalding, 1990). However, later offers made through legal/official channels are believed to have been turned down.

Mexico. Despite a ban on waste imports (with the exception of wastes bound for recycling), Mexico has often been a target for the waste trade, especially along the US–Mexico border. Several of the shipments of toxic wastes come from the US into Mexico, and are generally meant for recycling and disposal (Vallette and Spalding, 1990). Indeed, a 1986 US–Mexican treaty regulates the shipment of hazardous wastes between the two countries, with Mexico accepting hazardous wastes destined for recycling, but not for disposal; nonetheless there is potential for abuse of the export process, and abuses are believed to have occurred by shipping wastes for disposal under the disguise of recycling. Also, between 1992 and 1993, a UK company (dismantling the Capper Pass metal smelter in Humberside) was reported to be sending tin slags to Mexico (and also Bolivia). A consignment of 500 tonnes of the wastes (out of the approximately 3500 tonnes of total wastes) was sent from the UK to Mexico. After a protest, the wastes were returned to the original site (Greenpeace, 1993).

Nigeria. In an incident in Nigeria, several drums containing PCB-mixed wastes were illegally dumped in the port of Koko between 1987 and 1988.

This shipment consisted of about 4000 tonnes of toxic waste shipped from Italy, more than 150 tonnes of which was believed to comprise PCBs (Third World Network, 1989; Vallette, 1989; Vir, 1989; CIR and Moyers, 1990; Vallette and Spalding, 1990). The wastes were stored at rented space/land at a cost of about $100 per month. This whole affair was reportedly the result of an illegal arrangement between some Italian businessmen and some corrupt Nigerian officials. Eventually the wastes that remained – following the leakage of some of the toxic chemicals into the environment – were sent back to Italy in a 'fightback' that the Nigerian government embarked on, to fight this so-called 'toxic terrorism' – as the whole affair came to be called.

Poland. Extensive illegal dumping is believed to have taken place in 1988. For example, an Austrian company (Industrie Abfall Verwertung Gesellschaft) apparently transported about 50 tonnes of toxic wastes to Kraków and Lódź, which were misleadingly marked as 'neutral stains and lacquers'. They were, however, an apparent mixture of different hazardous wastes. Also, in 1989, three loads of 20 tonnes each were shipped from Germany into Poland. The wastes were described as non-ferrous metal residues, and therefore no transfrontier shipping documentation accompanied them. This importation was discovered in 1990 and the case ended up in court, following which the wastes were returned to Germany (Greenpeace, 1993).

Sierra Leone. Sierra Leone reportedly received a shipment of toxic ash containing several heavy metals (including cadmium, chromium, lead and mercury) into Freetown from the US in March 1989 (Vir, 1989). Previously, some illegal dumpings in 1987 led to some arrests being made. Prior negotiated waste trade deals in 1980 had been cancelled due to protests. Other similar schemes are believed to have been rejected.

South Africa. It is believed that some US-based firms allegedly shipped PCB wastes to South Africa in 1980. Furthermore, it is noted that mercury-contaminated wastes have been exported from the US and Europe to South Africa each year since 1986, with annual shipments totalling about 14 tonnes (Vallette and Spalding, 1990). In fact, each year, British-owned Thor Chemicals imports large quantities of mercury wastes from the US and Europe, and burns them in its mercury reprocessing plant in South Africa. Several other chemical wastes are also believed to arrive in South Africa from the US and a number of European countries for incineration and/or disposal. It is noteworthy that, whereas most of Africa became, in principle, completely off-limits to waste dumping, South Africa continued to offer an open door to the international waste traders/brokers. Apparently the South African government has no legislation prohibiting waste imports; not surprisingly, therefore, they had not attempted to halt waste shipments to Thor Chemicals' mercury reprocessing plant.

Spain. In 1990 thousands of mercury-contaminated waste was found at a dump site in the district of Almaden; the case involved the dumping, since 1980, of about 12 000 tonnes of hazardous wastes from chlor-alkali, battery and pesticide manufacture in France, Germany, Italy, the Netherlands, Norway, Sweden, Switzerland and the UK. Also, in 1988 and 1989, about 2000 tonnes of wastes were buried in a landfill at San Fernando de Henares and in 1991, as a result of an order from the Spanish government, the waste were removed and buried in a new landfill.

Turkey. In 1988 some radioactive material was apparently imported into Turkey. This consisted of about 1850 tonnes of radioactive waste materials originating from Germany. Subsequently the government of Turkey issued a legislative ruling and regulation for the strict control of all imported raw materials.

UK. The UK has featured prominently on the list of importers of hazardous wastes. In fact, there was an apparently large increase of waste imports into the UK from 1980 to 1988. This happened due in part to the increasing stringency of regulations in continental Europe and North America. As a result, waste imports increased from 5000 tonnes in 1983 to 185 000 tonnes in 1987, and about 80 000 tonnes of so-called 'special wastes' were imported in 1988. For example, between 1985 and 1986, about 28 000 tonnes of wastes were imported for incineration and chemical neutralization; between 1986 and 1987 the amount imported was about 60 000 tonnes; and in 1989 it was not considered unusual when several shipments of hazardous wastes that contained organic solvents, acid, alkalis and heavy metals were imported from Germany. In any case, the overall waste shipments dropped to about 40 000 tonnes of imported hazardous waste in 1989, following the implementation of the EC Directive on transfrontier movements of hazardous wastes. In general the predominant part of the imports originated from European countries where the costs of dumping of hazardous wastes are about five times greater that in the UK. About half of these imported wastes had come from the Netherlands, but Ireland, Belgium and thirteen other European countries add to the list of exporters.

Zimbabwe. In 1983–84 some waste brokers in the US set up a deceptive contract to ship virgin dry-cleaning chemicals and degreasing solvents to Zimbabwe, when infact, they were knowingly shipping toxic chlorinated waste mixtures – the transaction having been executed by falsifying the accompanying documents for the shipments (CIR and Moyers, 1990; Vallette and Spalding, 1990). Interestingly, financing for this programme had been provided by the US government; as a consequence, an investigation was conducted to bring the waste brokers to justice, but not until after the waste materials had arrived in Zimbabwe and been disposed of down an abandoned mine shaft.

Several other developing countries (including the Bahamas, Bangladesh, Belize, Bermuda, Gambia, Indonesia, Niger, Panama, Senegal, Sudan, Sri Lanka, Togo and Zambia) have apparently been approached with trade offers in hazardous wastes which were all rejected (Vir, 1989; Vallette and Spalding, 1990). In the meantime, many developing countries have, in principle, banned waste imports; thus, any waste trade deals are bound to be considered illegal and will therefore be annulled – even if achievable through diplomatic channels. On the other hand, whereas many less-industrialized countries have banned waste imports, many of the Eastern European countries (e.g. Poland has actively participated in several past waste trade schemes) and China remain vulnerable and legally open to international hazardous waste trade schemes.

Compared to other developing regions of the world, the Middle East and Mediterranean regions have been less of a target for waste exports in the past, but are by no means immune to international waste trade schemes. The ultimate responsibility of controlling or even eliminating waste trade activities apparently lies with the countries and industry generating the wastes. It is noteworthy that although the US may be the largest producer of hazardous wastes among all the industrial nations, it has not been alone in any extensive involvement in the waste trade – be it legal or illegal. In fact, throughout the 1980s Italy had even greater involvement in the waste export schemes undertaken by the industrial nations. In June 1989, however, Italy adopted Europe's most prohibitive waste export law which legally bans wastes from being exported from Italy to any non-OECD country.

4.2.2 MULTILATERAL WASTE TRADE ARRANGEMENTS

Industrialized nations, such as Canada and the US in particular, have over the years built flourishing businesses in the transboundary waste shipments that are tantamount to a waste trade. Canada is a major importer and exporter of hazardous wastes; large quantities of wastes are shipped consistently between Canada and the US under a bilateral treaty. According to a US EPA estimate, Canada is believed to receive about 90% of all wastes 'legally' exported from the US for disposal, primarily in an incinerator in Ontario and a landfill near Montreal, Quebec; similarly, Canadian wastes account for virtually all the wastes imported by the US (Vallette and Spalding, 1990). The Ontario Ministry of Environment estimates that about 150 000 tonnes of toxic wastes enter Canada from the US annually, while about 50 000 tonnes leave Canada for disposal in the US (CIR and Moyers, 1990; Vallette and Spalding, 1990). Also shipments (mostly PCB wastes) have been made from Canada to the UK.

Similar movements of wastes take place within most of the European communities, some of which have caused uproar among environmental organizations; for instance, Switzerland has been accused of exporting 'more toxic wastes than cheese or chocolate!' (Greenpeace, 1993). The UK had been a major importer of PCB wastes from throughout the world for several years; Canada and several other countries have therefore exported large quantities of PCB wastes to the UK every year, but this is apparently tailing down due to public pressures on both the exporting and importing countries. Both Australia and New Zealand have also shipped PCB-contaminated wastes to an incineration facility in the UK, but a recent UK decision to ban waste trade activities with Australia has put an end to this.

Indeed the UK has been a major recipient of hazardous wastes from over 20 countries throughout the world, but mostly from Europe. Overall, however, France has by far been the biggest importer of wastes in the EU, and former West Germany had been the largest exporter of hazardous wastes in Europe. Austria also is one of the major exporters of hazardous wastes to other European countries, but is authorized under relatively strict conditions.

4.2.3 WASTE EXCHANGE PROGRAMMES

Waste exchange programmes probably represent the most desirable forms of waste trade schemes yet. It is a fact, that throughout most of Europe and North America, there exist efficiently operated networks of hazardous waste exchanges. These can be likened to clearing-houses that coordinate operations between buyers and sellers of industrial by-products that are to be used in the production cycle again. Specific benefits to such programmes include potential savings from disposal costs, possible savings on the costs of virgin raw materials, and a lessening of the quantities of by-products of production that may legally be considered as hazardous.

The first hazardous waste exchange programmes in Europe began their operations in the Netherlands in 1969. These were initiated by trade associations, and technological or research institutes of the various chambers of commerce. Following this, the national chambers of commerce, in conjunction with various regional chamber of commerce offices, have been participating in hazardous waste exchange operations in different European countries. As an example, a unique arrangement has been established in Scandinavia in which the different participating companies maintain hazardous waste exchanges within and between Denmark, Norway, Sweden, Finland and Iceland. An umbrella-type organization, known as the Nordic Waste Exchange serves as a coordinating body for the exchange of information in all of the Scandinavian countries, and over 12 000 firms are potentially served by its operation (Crepeau and Beltz, 1985). Another operating network of waste exchanges exists in Germany. The German operations began in 1974, and most local and regional chambers of commerce have been collecting information from industries within their area on the availability of production by-products. A special office serves as a conduit for the dissemination of lists of available production residues, which are then published in a monthly bulletin and can be accessed by all local or regional offices. Similar programmes, even if less sophisticated, exist in the North American region (i.e. the US and Canada).

Further development of the European hazardous waste exchanges lies in the cross-frontier network of exchanges maintained by the German National Chamber of Commerce and Industry. The Nordic Waste Exchange, as well as similar exchanges in Austria, Switzerland, Italy, Luxembourg and France periodically forward lists of production residues needed by firms to the Bonn, Germany office. In fact, in 1981 alone, over 2600 offers and inquiries were handled by the German office which, in turn, forwarded the lists to all countries participating in the network. The German Waste Exchange Office publishes in its quarterly bulletin, a dictionary of the more important terminology involved in the programme; these are presented in German, English, French and Italian.

It is apparent that hazardous waste exchange programmes will continue to be in operation for as long as there is a need in the industrial community. The special features of hazardous waste exchange in Europe include the almost exclusive initial funding and continued maintenance by the industries which the waste exchange programmes serve. The chambers of commerce are the primary vehicles of such funding.

4.3 PUBLIC OUTCRY, INDECISIONS AND UNCERTAINTIES ABOUT THE FUTURE

Traditionally in places like the US, hazardous waste TSDFs have been disproportionately located in poor and minority communities. Lately, however, there is resistance against what has come to be branded as 'environmental racism'. Consequently the developing and newly industrializing countries seem to have become more important targets for many an industrialized nation. Waste shipments abroad have indeed been on the rise globally. In the US alone, notices of intent to export hazardous wastes to foreign countries that were issued by the US EPA – as required by US law – rose dramatically (Figure 4.2) from a mere 12 in 1980 to 258 in 1986, then to 570 in 1988, and to about 626 in 1989 (Opsal, 1989; CIR and Moyers, 1990; Vallette and Spalding, 1990). But new international agreements designed to help abate this trend promise to offer new challenges.

In attempts to avert public confrontations associated with the waste export schemes, a new unresolved and less-challenged problem has been born. Now, wastes are being shipped to vulnerable countries in the name of recycling. But again, environmentalists and other responsible watch-

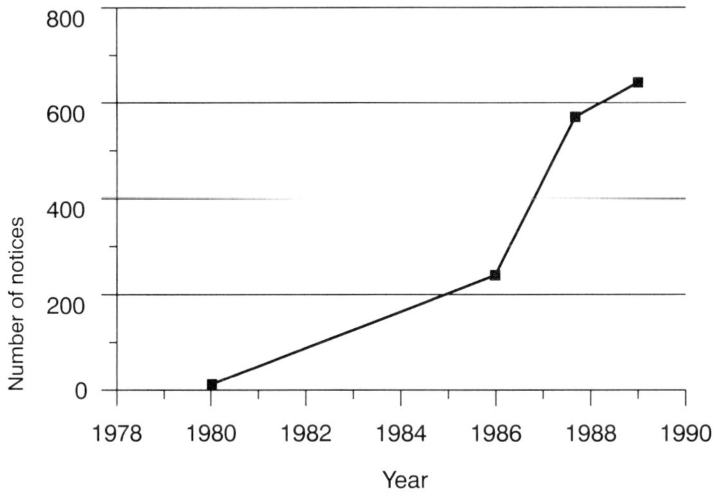

Fig. 4.2 Growth trend for notices of intent to export hazardous wastes to foreign countries, issued by the US EPA between 1980 and 1989.

dogs perceive the dangers existing for the developing countries as a result of materials and industrial by-products or wastes being shipped into such countries as 'recyclables'; some of the materials meant for recycling may contain several constituents (such as components of mercury residue, lead-acid batteries, etc.) from which valuable materials are expected to be extracted using cheap labour in the recipient developing countries, and where people usually work under unhealthy and 'unacceptable' conditions. Furthermore, nearby residents to industrial plants where such activities take place are also threatened, and so is the general ecological system in these regions. This scenario therefore presents yet another challenge to the continuing trade in hazardous wastes that could be going on in the name of a 'sham' recycling programme.

In fact, lately, several developing countries have been targeted as potential recipients of wastes from certain industrialized countries. The major difference in the waste trade between the industrialized countries as compared with exports to the developing countries is that such wastes are disposed of in a relatively safe manner in the former, compared with what happens when the wastes find their way into the developing countries. Notwithstanding the apparent high-technology waste management facilities and processes utilized in the industrialized countries, these are not without potential problems even under all the controls of an 'eagle's eye'. This means that even in technologically advanced nations, trading and/or dumping of wastes has the potential to jeopardize human health and the environment.

All of the above indications clearly demonstrate and point to the notion of, and the application of, the concept of double standards (i.e. a situation when one performance level is used for one's own country or region and another for others/elsewhere, thus offering different degrees of protection for the dichotomized societies) in waste trade activities. However, the apparent double standards associated with the waste export schemes were also bound to raise protests from different quarters of the international community. This is reflected, for instance, in a declaration by the Nigerian president in May 1988 at the Organization of African Unity (OAU) summit in Addis Ababa, Ethiopia, that: 'No government, no matter the financial inducement, has the right to mortgage the destiny of future generations of African children', by participating in the toxic waste trade. Furthermore, several activist groups have taken very firm positions against the shipment of wastes to less developed countries – as reflected in a Greenpeace activist, Jim Vallette's declaration on the toxic waste trade as being 'morally reprehensible to force developing countries into this choice between poison and poverty' (Newsweek, 1988).

4.3.1 IS A *DE FACTO* DOUBLE STANDARD SUSTAINABLE?

If indeed all countries had the same level of environmental consciousness, or even the same general environmental standards, it is apparent that the choice of risks on which to formulate policy decisions may have to differ widely – if only due to differing climatic and other geographical or demographic conditions. For such reasons, even in a world where all individuals or communities shared the same environmental preferences/ideals, environmental laws and regulations would still be likely to differ substantially between countries; and even where they were the same, the costs to industry of complying with those laws and regulations would still be likely to differ substantially from country to country (Trebilcock and Howse, 1995). Such differences can therefore encourage interest in utilizing what can be perceived as a comparative advantage in development opportunities.

Notwithstanding the preceding argument, in discussing the issue of double standards, Kerton (1990) notes that no economic activity can have a sustainable future if it is based on exposing the ill-informed or underprivileged to hazards, leaving them to possibly discover later that they have been badly or unfairly treated. Consequently, both human welfare and integrity require that double standards be eliminated, especially insofar as they relates to the waste trade. Indeed it is in nobody's interest to ignore the problem of double standards because to abuse them – as has so often occurred with waste trade schemes – can foster cynicism, distrust and fear (Kerton, 1990). In general, public welfare is best served when individual citizens and decision-makers clearly understand the true risks and benefits of a given activity; unfortunately, such information is often withheld, concealed or misapplied in most cases/situations.

Of course, double standards may be justified in very rare and selective cases of human activities, but generally not in situations or forums such as is presented by waste trade activities. A double standard may be appropriate where an activity offers a mix of benefits and risks, and where operating conditions or environments differ enough to justify different risk-benefit outcomes; but most contemporary double standards cannot be reasonably justified (Kerton, 1990). In fact, the export of hazardous wastes from developed nations to developing countries, or indeed to any country with less efficient waste management and disposal practices (namely, disposal technologies and environmental infrastructure) than the exporting country, may be considered as constituting a *de facto* double standard; this is because the activity will generally result in lesser environmental and human health protection for the importing country compared with the waste exporting country. The notion that a *de facto* double standard could result in sustainable development may therefore be seen only as an illusion, and thus should be considered unacceptable for most waste trade situations.

4.4 THE FALLACIES AND REALITIES OF THE INTERNATIONAL WASTE TRADE

The NIMBY (not-in-my-backyard), NIABY (not-in-anyone's-backyard), NIMTO (not-in-my-term-of-office), LULU (locally-unwanted-land-use), and related syndromes seem to be catching up with many nations that may become unwilling partners to the hazardous waste trade in the foreseeable future. However, this alone may not be the answer to curbing the waste trade problem – unless adequate and implementable environmental control measures are put in place internationally. This is because, short of effective international monitoring machinery or mechanisms, multinational corporations may decide to relocate some 'dirty' industries to the developing nations where they believe environmental laws are less stringent. Indeed, in a way, this has indirectly happened in Mexico under the so-called Maquiladora Program.

The Maquiladora Program was implemented in 1965 as part of Mexico's Border Industrialization Program, designed to generate employment and stimulate industry using the abundant low-cost labour force in Mexico. Maquiladora operations utilize local labour to manufacture or rework materials and components imported into the country on a temporary basis; raw materials for such industries enter the country duty-free. Products from the operations are exported, with tariff payments being only on the value-added component due to the processing in the country. Under the provisions of the Maquiladora Decree, any hazardous wastes generated during the Maquiladora processing must go back to the country of origin of the raw materials, although certain recyclable wastes may stay behind (US/Mexico Hazardous Waste Work Group, 1989; Merk, 1990; Shimmin, 1990; Harbison and McLarty, 1996).

However, Mexico now recognizes the environmental deterioration that has resulted from the explosive growth of the Maquiladora Program and improper management of industrial wastes. Wastes from the Maquiladoras is supposed to be shipped back to the country of origin of the raw materials – but this does **not** seem to happen on any consistent basis. Indeed it is feared that untreated hazardous wastes have been dumped in several locations in the proximity of the US–Mexico border (especially in the poor Mexican neighbourhoods). The problem has also been compounded by the lack of adequate legal and regulatory systems to protect public health and the environment. To counteract this problem, the Mexican government subsequently enacted much stronger and enforceable legal regulations that directly affect the Maquiladora industry. Severe liabilty tags (including criminal sanctions) are attached to the new regulations. Furthermore, in the period leading up to the negotiation and ratification of the North American Free Trade Agreement (NAFTA), Mexican officials began more intensive and serious enforcement of the nation's environmental regulations. In fact, in 1992, in the context of the NAFTA negotiations, the US and Mexico stated their joint intention, in the 'Border Environmental Plan', to work together on transboundary pollution problems (Ginn, in Avogadro and Ragaini, 1994; Kublicki, 1994; Harbison and McLarty, 1996). Even so, the world is likely to see more situations like this, whereby the industrialized nations take advantage of both cheap labour and less stringent environmental laws to relocate some 'dirty' industries in, or diversify into developing and/or newly industrializing countries. Unfortunately only a few such countries with an emerging economy have the basic infrastructure and services needed for the safe disposal of industrial wastes (Leonard, 1985; Harbison and McLarty, 1996). Most of the countries with emerging economies or economies in transition may therefore tend to rely on the natural assimilative capacity of natural systems, which will invariably become overloaded and overwhelmed by the sheer quantities of wastes released into the various environmental compartments.

Ultimately even if the international community succeeded in implementing a total ban on direct waste trade activities, developing and newly industrializing countries will remain vulnerable to related activities. One area of major concern is the mushrooming of recycling plants in some developing countries, particularly lead recycling plants (as found in Brazil, Taiwan, Mexico, India, China, South Korea, South Africa, etc.). Although used batteries contain lead and acid, current environmental regulations in countries like the US do exempt intact batteries from being classified as hazardous waste and can therefore easily be exported for recycling; and industry estimates that at least 70 million automobile batteries – containing sulphuric acid and lead (known to cause damage to the central nervous system (CNS) in children, among several other ailments) – will be discarded in the US alone each year during the 1990s (CIR and Moyers, 1990). Indeed, contamination from lead smelters in Taiwan has been determined to be so extensive that the Taiwan Environmental Protection Administration decided to end all used battery imports by 1990, and to further upgrade the smelters to reprocess its own batteries safely. But then, just when Taiwan began closing some of its doors to the international trafficking of wastes (including the importation of used lead-acid batteries), neighbouring China opened its doors to the same problems and potential waste streams in a similar trade practice. The fear is some other developing countries may find themselves importing wastes in the name of a possible 'sham' recycling. In fact, until effective global machinery is in place to control and manage the global waste problems, populations in these countries will remain vulnerable to the effects of potential exposures to hazardous wastes.

Meanwhile there is continuing public outcry, restraints and legislative initiatives towards the hazardous waste trade. The reaction of the developing countries to the hazardous waste trade appears to be strong. But the 'billion-dollar' question is whether these antagonisms will indeed halt this trade practice. Is this opposition any different from the drug-trade-cum-drug-war in which more questions than answers still exist to date? People with criminal intentions can reap huge financial returns with minimum effort and are therefore willing to bear the risks involved. After all, for the criminals, this may not be any different from making high-risk investments on the stock-exchange/money-markets, in the hope for high returns, and the game goes on despite the potential high losses that they may incur if something went wrong.

Under the prevailing circumstances, most developing countries will not be immune to waste trade activities for the foreseeable future. In fact, enforcement is difficult and possible corrupt practices by some local officials could create many opportunities for circumventing any restrictions. Also, given the potential for high profits, it is likely that the market for hazardous wastes in developing and newly industrializing countries will continue to be profitable and available – as demonstrated, for instance, by the decision by China to open its doors to what Taiwan had shut off: the importation of used car batteries.

Controlling the transboundary movements of hazardous wastes

5

Laws controlling hazardous waste disposal are now energetically enforced in most industrialized countries. As these waste control laws get tighter, industries have come to assume greater financial responsibility and are trying to lessen this burden by finding alternative methods and places for waste disposal. Some have exported wastes to developing countries, where laws are generally less stringent (if they exist at all) and/or less strictly enforced (where they exist). Indeed, to a great extent, it is the advancement of environmental standards in most industrialized countries that has led to the shipment of toxic wastes to other countries lacking stringency in their laws, or to jurisdictions that may not be fully aware of the implications of accepting hazardous wastes to their regions. In particular, the transfrontier movements of hazardous wastes between developed and developing countries are significantly influenced by regulatory standards and requirements – or lack thereof. The harmonization of hazardous waste management standards across all nations, irrespective of the level of development, will therefore be necessary, if measurable results are to be achieved with regard to optimizing the use of local disposal options and controlling the transfrontier movements of hazardous wastes.

In the 1980s, there were more than one hundred reported instances of proposals for transfer of potentially hazardous wastes from OECD Member countries to developing countries (Greenpeace, 1988; 1994). Over half of these proposals have been put forward since the beginning of 1986; more than 40 developing countries were reportedly targeted to receive such wastes. Eventually there were outcries against such waste traffic, as exemplified by the following.

- The OAU unanimously passed a resolution condemning the dumping of potentially hazardous wastes on African soil.
- The European parliament passed a resolution condemning all significant exports of potentially hazardous wastes from the EC to any country with a developing economy.
- The Commission of European Communities negotiated a ban on exports of hazardous wastes from the then 12 Member States to non-Member States who are party to the Lome IV Convention.
- Legislation has been introduced in the US Congress, which would prohibit the export of any wastes (except for certain recyclable materials) unless a special permit (export licence) were granted stipulating that the wastes would be disposed as if it were in the US.

In addition, individual countries have restricted or prohibited imports of wastes or hazardous wastes. For example, the Dominican Republic has long (since 1984) prohibited the import of industrial toxic wastes, and several dozen more countries had declared their territories off-limits to foreign wastes by mid-1989.

In fact, transfrontier movements of hazardous wastes have lately been under close public scrutiny, resulting in the accession of several international agreements to regulate such activities. The most significant control measures and major agreements are discussed in the proceeding sections in this chapter.

5.1 OECD AND EU INITIATIVES CONCERNING THE TRANSFRONTIER MOVEMENTS OF HAZARDOUS WASTES

In the beginning of the 1980s many OECD countries had adopted or were in the process of developing regulatory measures to enable their authorities to monitor the management of hazardous wastes from the place of generation to the place of disposal. It became rapidly clear, however, that in the case of transfrontier movements of wastes, such national monitoring systems were not totally adequate since countries had generally insufficient knowledge about consignments of wastes imported into their territory to exercise proper control.

In 1982 the Waste Management Policy Group (which was originally established in 1974 by the Environment Committee) proposed that guidelines be developed for the export and import of hazardous wastes bearing in mind the different levels of development of environmental regulations among countries and the different levels of expertise.

In 1984 the OECD Council decided that: 'Member countries shall control the transfrontier movements of hazardous wastes and, for this purpose, shall ensure that the competent authorities of the countries concerned are provided with adequate and timely information concerning such movements.' Moreover, a comprehensive set of guiding principles concerning such control was recommended to Member country governments. The principles included with this OECD Council Decision and Recommendation related to the basic policy strategies needed in order to properly monitor and control international traffic in potentially hazardous wastes.

Furthermore, in 1984 the OECD issued recommendations on the exchange of information when exporting chemicals that have been banned or severely restricted. This became a starting-point for developing methods to control the transboundary movements of waste materials. In the EU countries, the export of wastes is controlled under Council Regulation 93/259/EEC, which bans all exports of wastes for final disposal outside the Union and the European Free Trade Association (EFTA). There are also specific controls on hazardous wastes being exported for recovery, and such exports from the EU, other than to OECD countries, are to be banned after 1997 as a consequence of a decision taken at the Second Conference of Parties in 1994.

A 1985 ministerial level meeting of the OECD Environment Committee declared that Member country governments would: 'strengthen control and the monitoring of the generation and disposal of hazardous wastes and establish an effective and legally binding system for control of their transfrontier movements, including movements to non-Member countries'. The 'monitoring' means that the whereabouts of hazardous wastes are known 'from cradle-to-grave', and that the wastes are directed to an appropriate facility for treatment and disposal; 'control' means that authorities are aware of waste flows, and can act rapidly to ensure that the possibility for inappropriate handling of the wastes is minimized. This declaration was transformed into an action programme when, by means of Resolution C/85/100, the OECD Council decided to: 'develop an international system for effective control of transfrontier movements of hazardous wastes, which will include appropriate OECD instruments such as further Acts of the Council covering notification, identification and control of such transfrontier movements, as well as an international agreement of a legally binding character'. The OECD Council specifically instructed the Environment Committee to:

- undertake the work necessary to implement this Resolution, with a view to making proposals in the form of appropriate OECD instruments and a draft international agreement;
- assess whether the agreement should be developed in such a way that it can be open to both OECD Member countries and other interested countries;
- base the international system upon the principles contained in the Decision and Recommendation of the Council on Transfrontier

Movements of Hazardous Waste (C/83/180/Final), further developed in the light of the Conclusions and Recommendations adopted by the OECD Conference on International Cooperation Concerning Transfrontier Movements of Hazardous Wastes, and subsequently endorsed by the Environment Ministers of OECD Member countries.

The establishment of an OECD system to monitor and control transfrontier movements of hazardous wastes implies internal obligations for monitoring, control and enforcement on the part of OECD Member countries participating in the system; also, courses of actions that are undertaken for implementing any such system should not relieve the generator of hazardous wastes from any of its responsibilities. In any case, this system was to address issues concerning: the definition and classification of hazardous wastes; the notification, identification and control of the hazardous wastes; relations with non-Member countries; and the legal and regulatory framework. The system should also recognize and implement the principle that OECD Member countries would not apply any less strict controls on transfrontier movements of hazardous wastes involving non-Member countries than they would for movements involving only Member countries; and that they would not allow movements of hazardous wastes to non-Member countries to occur without the consent of the appropriate authorities of the importing country and of any non-Member countries of transit, and unless the hazardous wastes were directed to adequate disposal facilities in the importing country.

5.1.1 THE OECD CORE LIST OF HAZARDOUS WASTES

The identification and classification of hazardous wastes have yielded lists of potentially hazardous wastes in more than 20 OECD Member countries, with no two of these lists being identical. The existence of these diverse lists prompted Member country governments to explicitly call for the development of 'an agreed list of hazardous wastes for all transfrontier movements of such wastes between OECD Member countries', as part of Council Resolution C(85)100. This process indicated that wastes containing some proportion of one or more of 27 constituents were proscribed by all of the national and international lists. Certain types of generic wastes are often legally defined to be hazardous to human health and/or the environment.

Consistent with the OECD deliberation, certain countries adopted or considered the adoption of a national list of hazardous wastes following a 1988 proposal of the European Communities. The Commission of the European Communities proposed two new directives which would define 'wastes', 'disposal', and 'hazardous wastes' within the EU, and these were subsequently adopted.

Eventually negotiations among representatives of OECD Member countries resulted in adoption by the OECD Council in 1988, a decision [(88)90(Final)] that includes a Core List of wastes for which consensus was reached that they require control when proposed for disposal following transfrontier movements. As a result, one may predict with reasonable confidence that between 85 to 90% of all wastes legally defined as or considered to be hazardous by OECD Member countries and the EC, and over 90% of wastes that are prohibited or restricted from disposal at sea, are included in the Core List (OECD, 1993).

It is very important to note that there are no concentration limits or levels included with the Core List. What decides if a waste listed in the Core List is subject to control is whether that waste exhibits one or more characteristics which could cause harm to human health and/or the environment. These characteristics are listed in Council Decision C(88)90(Final) along with descriptions allowing a conclusion to be reached concerning the likelihood that a given amount of waste does or does not possess such characteristics as listed. A caveat included with the listing alludes to the fact that 'absolute' conclusions concerning whether a waste presents a hazard and, if so, what the degree of hazard might be are difficult to reach, given the present level of knowledge. The conclusion then, is that determination of concentration limits, which implies both qualitative and quantitative analysis of wastes, would be time consuming and expensive. The Core List coverage is such that control is flexible; wastes which do not exhibit a hazardous characteristic can be excluded. Hence the decision not to incorporate strict concentration thresholds should shield the system against unnecessary 'over-control' while seeking to ensure that hazardous wastes crossing national frontiers are indeed subject to suitable control measures (OECD, 1993).

In addition to wastes covered by the Core List, all other wastes that are considered to be or are legally defined as hazardous wastes are also subject to control under the terms of the OECD Council Decision C(88)90(Final). When this Decision was adopted, however, the Council agreed that when implementing this part of the Decision, 'Member countries shall not be obliged to enforce laws other than their own.' Full coverage for countries involved in a transfrontier movement of hazardous wastes is, therefore, obtained by use of the Core List, plus any wastes legally defined as, or considered to be hazardous in the importing or exporting countries.

The Council Decision also provides for a definition of the terms 'wastes' and 'disposal' for purposes of control of transfrontier movements. In practical terms, 'wastes' are any materials which are subject to any of the 'disposal' operations. But any 'wastes' destined for operations which may lead to resource recovery, recycling, reclamation, direct reuse or alternative uses are only deemed to be 'hazardous wastes' if they are considered to be or are legally defined as hazardous wastes in either the country where they are generated or are to be disposed of or both (OECD, 1993).

Ultimately there is a real need for the introduction of a comprehensive International Waste Identification Code (IWIC), because many countries have differing definitions, lists and methods for describing wastes considered to be, or legally defined as hazardous wastes. The Final OECD Council Decision provides for such a uniform classification system. This allows virtually all wastes deemed to be hazardous by most countries to be described satisfactorily in terms of potential hazard, activity generating the wastes, physical form, generic descriptor, constituents, and an indication of the reasons why the materials were intended for disposal, together with the disposal operation to which the wastes will be subjected. The IWIC should provide a coded 'cradle-to-grave' dossier which fully describes any batch of wastes undergoing transfrontier movement. If it is included within a notification system for exports of hazardous wastes, the coding scheme of the IWIC will make the system more amenable to modern communication and information transfer technologies as well as to computer-based statistical evaluation.

5.1.2 DEVELOPMENT OF AN OECD SYSTEM FOR THE MONITORING AND CONTROL OF TRANSFRONTIER SHIPMENTS OF HAZARDOUS WASTES

A number of OECD Member countries have acted to monitor and control transfrontier movements of hazardous wastes by means of domestic legislation and regulations, as well as via bilateral and multilateral agreements. In this regard the EC adopted a series of directives aimed at monitoring and controlling transfrontier movements of hazardous wastes. For instance, Directive 84/631/EEC on the supervision and control of the transfrontier shipment of hazardous wastes within the EC was adopted in 1984; this Directive largely transposed the 'Principles' included within OECD Council Decision-Recommendation C(83)180(Final) into a legally binding form for use by the EC Member countries. The Directive excludes controls for chlorinated and organic solvents, and provides for minimal controls with respect to waste materials, scrap, sludge, ash and dust from non-ferrous metals which are intended for reuse, regeneration or recycling on the basis of a contractual agreement regarding such operations.

Specific requirements necessary for the notification, identification and control of the transfrontier movement of hazardous wastes typically will include the following.

- Those wastes that are referred to as 'hazardous wastes' for purposes of the control system, should be identified. Persons who are technically capable of correctly determining if a 'waste' is or is not 'hazardous', subject to the terms of the control system, carry the responsibility for such definitions.

- Authorities who are competent to judge whether a proposed import of hazardous wastes can be managed in accordance with the regulations and practices of the importing country must receive prior notification that such an import is planned. These authorities from the importing country must formally confirm that they have no objection to a proposed shipment of hazardous wastes into a territory under their jurisdiction.
- Simplified procedures should possibly be evaluated for wastes destined for recycling.
- All countries through which hazardous wastes are to transit must receive prior notification. If an objection is formally lodged by the authorities of a transit country to the passage of hazardous wastes, then the transfrontier movement cannot legally pass through this country until the objection is lifted.
- For transfrontier movements of hazardous wastes, only transporters and disposal firms who are authorized or allowed to carry out such operations may be employed or contracted.
- If a transfrontier movement of hazardous wastes cannot be completed as foreseen, the exporter must re-assume responsibility for its wastes, and efforts to ensure that the wastes are disposed of in an environmentally sound manner must be undertaken.
- Control of transfrontier movements of hazardous wastes for countries which are not parties to existing agreements should be no less protective of their environments and public health than control of such movements among contracting parties.
- Bilateral or multilateral agreements or arrangements should be allowed/permitted, provided that any such agreements or arrangements are compatible with the principles of any existing broader international agreements.
- Provisions should be made to provide for regular information exchange concerning all aspects of the implementation and functioning of any agreement on the cross-boundary waste movement programme.

Ultimately an appropriate monitoring and control system (whether an OECD system or any other for that matter) can be effective if and only if all parties understand and agree on what materials are deemed to be 'wastes' and also what 'wastes' are considered as 'hazardous wastes' and are subject to controls imposed by the system.

5.1.3 OECD LEGAL REQUIREMENTS AND ACTIONS CONCERNING THE EXPORT OF HAZARDOUS WASTES FROM THE OECD AREA

The OECD Council Decision of 1986 [(86)64(Final)] stipulates that Member countries must apply no less strict controls on transfrontier movements of hazardous wastes involving non-Member countries than would be applied if such movements involved only Member countries. The consequential obligations placed upon Member country governments require them to see to the following action items.

- Monitor and control exports of hazardous wastes to a final destination which is outside the OECD area.
- Ensure that their competent authorities have the legal power to prohibit these exports in appropriate (or rather, 'inappropriate') instances.
- Prohibit movements of hazardous wastes to a final destination in a non-Member country without the consent of that country.
- Ensure that any countries of transit are notified in advance of the proposed transfrontier movements.
- Prohibit movements of hazardous wastes to a non-Member country unless the wastes are directed to an 'adequate' disposal facility in that country.

Decisions of the OECD Council are legally binding on Member countries which vote for them. For the OECD Member countries, Directive 86/279/EEC of 1986 sets forth precisely these same requirements concerning control of exports of hazardous wastes outside the EC.

Since 1988, the legally binding obligations placed upon OECD Member countries with respect to exports of hazardous wastes to non-Member countries – especially to developing countries – have included the following stipulations (OECD, 1993).

- Competent authorities of the exporting country must be notified of all proposals to export hazardous wastes delineated by OECD Council Decision C(88)90(Final) prior to the commencement and the movement.
- The competent authorities must object to waste shipments if,
  (a) the importing country has not been notified and if consent of an appropriate official in the importing country is not available;
  (b) the wastes are not directed towards an adequate disposal facility; and/or
  (c) proof that all transit countries have been notified of the prospective movement is not available.
- The competent authorities must object to any proposal to export the specified wastes to any country that have banned the import of hazardous wastes, since the consent of the importing country cannot be obtained legally.
- The wastes must be appropriately identified and classified.
- The wastes must be packaged, labelled, documented and carried according to any applicable international agreements governing the transport of dangerous goods.
- The shipment cannot legally occur until the competent authorities inform the exporter that the proposed transfrontier movement can proceed.

It is interesting to note, however, that if certain wastes included in the Core List are to be exported from a Member country to a non-Member country for purposes of one or more of the specific recycling/reclamation operations indicated by the Decision document, and if these 'wastes' are not legally defined as, or considered to be hazardous wastes in the country of export, then the requirements of Council Decision C(86)64(Final) do not apply. Also the term 'disposal' is defined by Council Decision C(88)90(Final) but the terms 'adequate disposal facility' are not specifically defined. As a result, Member countries can exercise considerable flexibility in reaching conclusions concerning whether a proposed export of hazardous wastes will be directed to an adequate disposal facility.

Finally, it is apparent that well-established uniform rules for packaging, labelling and transporting dangerous goods currently exist. These rules should be enforced with respect to transfrontier movements of hazardous wastes from Member countries to places both inside and outside the OECD area.

5.2 BILATERAL AGREEMENTS ON THE TRANSBOUNDARY MOVEMENTS OF HAZARDOUS MATERIALS IN NORTH AMERICA

The second half of the 1980s saw the US concluding two bilateral agreements on the transboundary movements of hazardous wastes with Canada and Mexico. The original 'Agreement between the Government of Canada and the Government of the United States of America Concerning the Transboundary Movement of Hazardous Waste' was adopted in Ottawa on 28 October 1986; the 'Agreement on Co-operation between the United States of America and the United Mexican States regarding the Transboundary Shipments of Hazardous Wastes and Hazardous Substances (Annex III to the 1983 Agreement between the USA and Mexico on Co-operation for the Protection and Improvement of the Environment in the Border Area)' was adopted in Washington DC on 12 November 1986. These agreements have since been amended, especially to cover certain provisions of NAFTA between Canada, Mexico and the US.

The agreement between Canada and the US on transboundary movements of hazardous wastes is of especially significant interest, as most of the hazardous wastes legally exported from the US go to Canada. Significant provisions of this agreement include the following.

- Lays down the principles of hazardous waste movements taking place across the common borders of the two contracting parties.
- Scope of wastes covered is to be determined by the national legislation of the contracting parties.

- Provides for notification of exports of hazardous and other wastes from one country to or through the other, and adopts the concept of prior informed consent of the importing or transit state.
- Parties are required to cooperate in the implementation of the agreement, to include monitoring and spot-checking of waste shipments.
- Country of export must re-admit any waste shipment that has been refused by the importing country for any reason.

The parties to the agreement may consider enacting and adopting national regulations that may be deemed necessary to help in observance and implementation of the provisions of this agreement.

The bilateral agreement between the US and Mexico on the transboundary movements of hazardous wastes and hazardous substances covers both hazardous wastes and substances which are banned or severely restricted in either country. The general provisions of this agreement are similar to the ones enumerated above for the Canada–US agreement.

5.3 INITIATIVES BY THE UN TO CONTROL THE INTERNATIONAL TRADE IN HAZARDOUS WASTES

The subject of controlling the transnational shipments of hazardous materials was recognized as a very important one in the latter part of 1981, at the Montevideo Meeting of Senior Government Officials/Experts in Environmental Law. This meeting was convened to establish a framework, methods and programme (including global, regional and national efforts) for the development and periodic review of environmental law, and to contribute to the preparation and implementation of the environmental law component of the system-wide medium-term environment programme.

Subsequent to the 1981 Montevideo meeting, in 1987 the UNEP 1985 Cairo Guidelines and Principles of Environmentally Sound Management of Hazardous Wastes were adopted by the Governing Council of the UNEP, which also requested the development of a global convention on the control of transboundary movements of hazardous wastes. The UNEP Secretariat adopted a very ambitious schedule; a convention was to be proposed for adoption in 1989. In keeping with the instructions of the UNEP Governing Council's decisions of 1987, some portions of the convention are taken verbatim from, or are close paraphrases of, the earlier OECD draft international agreement. Ultimately this work contributed significantly to the preparation of the UNEP Global Convention adopted in Basel on 22 March 1989, and became known as the Basel Convention (see section 5.3.1).

In other related developments, the UNEP followed the OECD's lead, adopting in 1987 the London Guidelines for the Exchange of Information on Chemicals in International Trade; this instrument was later amended by adding a system of Prior Informed Consent (PIC), adopting in 1989 the Amended London Guidelines. In parallel, the Food and Agricultural Organization (FAO) in 1989 also amended its Code of Conduct on the Distribution and Use of Pesticides. The FAO and the UNEP are jointly implementing PIC in a voluntary global system to control trade in pesticides and chemicals. These two UN bodies have added a third category of dangerous materials, namely, substances withdrawn by companies from testing or consideration for domestic regulation due to health or environmental reason – the intention being to close a loophole that allowed firms to protect export markets when they have removed their dangerous products from domestic scrutiny. The PIC policy has been in operation since 1991 and applies to 12 pesticides and 5 industrial chemicals. As of 1995, 127 countries participated in the PIC system. A Decision Guidance Document (DGD) for each chemical or pesticide has been sent to all participating government agencies, which then make a decision about whether, and under what conditions, to allow future imports. It is noteworthy that the PIC system has revealed that many substances are hazardous only under specific conditions of use. National regulations might therefore allow substantial usage under such conditions. Regulations such as these often do not qualify as bans or severe restrictions. Consequently, the substances remain outside PIC rules and it is not easy to use the powerful trade provisions of PIC to shift enforcement from importers to exporters.

5.3.1 THE BASEL CONVENTION ON THE CONTROL OF TRANSBOUNDARY MOVEMENTS OF HAZARDOUS WASTES AND THEIR DISPOSAL

Any transboundary movements of wastes across national borders constitute a form of waste trade. In this light, a new kind of trade seems to be emerging, and the international community sees a need to regulate this activity. Consequently, in March 1989, ministerial level delegates from some 116 countries meeting in Basel, Switzerland, agreed on the need for a treaty for controlling the transboundary movements of hazardous wastes, but only 33 countries signed it then; the treaty set global rules of procedure for the international waste trade.

This treaty – the Basel Convention on the Control of Transboundary Movements of Hazardous Wastes and their Disposal, sponsored by the UNEP – is a global convention developed by the UNEP and intended to help minimize and control international trade in hazardous wastes. It establishes notification procedures to be used before a country may export hazardous wastes. The Convention adopted several aspects of earlier developments of the OECD on issues relating to the transfrontier movements of hazardous wastes (OECD, 1985a,b; 1986; 1988).

The Basel Convention represents a first step in defining the global means to reduce and strictly control the movements of hazardous wastes, and to ensure that these wastes are disposed of in an environmentally sound manner. It provides realistic measures to strengthen the protection of the global environment from the possible harmful effects of the transboundary movements of hazardous wastes and their disposal. The Convention focuses on the protection of human health and the environment. However, this Convention apparently disappointed many developing countries and environmental activist organizations, who had urged that a ban on waste exports from industrialized to less-industrialized countries be incorporated. It has been speculated by some that the Basel Convention leaves much room for subjective interpretation and thereby the circumvention of many of its provisions. During the early parts of the formulation of this agreement, Greenpeace had indicated that: 'The greatest danger ... of this Convention is that it creates the illusion that the international waste trade is now under control. This ... will do more to institutionalize waste trade than it will to prevent it' (Greenpeace, *Waste Trade Update*, Volume 2, Issue 3; July 15, 1989). Even so, one could argue that institutionalizing the hazardous waste trade may not by itself be undesirable, provided that the adopted guidelines fully protect the interests of the developing countries, and provided that applicable compensations are fair and adequate.

Overall, the Basel Convention calls for international cooperation between parties in the building and/or the improvement of national capabilities to manage hazardous wastes in an environmentally sound manner, as well as for the development of a technical and legal infrastructure (including the necessary legislation and regulations). The Convention also allocates to the exporter states, the responsibility for ensuring that exported waste is managed in a safe manner (UNEP, 1989). However, because of the apparent notion of institutionalizing an unacceptable and unfair programme, the less-industrialized countries had been reluctant to join the Basel Convention. In fact, at the Convention, the African nations present and most developing countries fought for an absolute ban on waste exports, and took the position that all such trade is exploitative and should be banned. On the other hand, most of the industrialized nations took an opposing stand, and, the proposal of the African countries was rejected. Nonetheless, what the Convention set out to do was to become a good starting-point for the global control of transfrontier movements of hazardous materials.

Radioactive materials were not covered under the Basel Convention, but have been addressed in a parallel agreement reached by the International Atomic Energy Agency (IAEA). At its May 1989 meeting in Vienna, Austria, the IAEA came up with guidelines that paralleled the Basel agreement, but specifically affected international trade in radioactive wastes. The subsequent IAEA code developed in February 1990 to control such trade attempts to govern the exports and disposal of radioactive wastes using a procedure similar to the Basel Convention.

5.3.1.1 General elements and provisions of the Basel Convention

The Basel Convention is a global environmental treaty that strictly controls the transboundary movements of hazardous wastes. The Convention places on exporting and importing states, a shared responsibility for environmentally sound management and disposal of traded wastes, while affirming the sovereign right of any state to prohibit imports. It primarily requires the notification and consent of recipient countries before any waste shipments can take place. Under this set of rules, the exporting signatory country must send notification to countries of import and transit, whether or not the latter are also signatories to the Convention. The party of export should not proceed with the export until written consent is given by the importing and transit countries involved (UNEP, 1989; Vallette and Spalding, 1990). Indeed the heart of the Convention is a requirement that importing countries are notified of, and consent to, waste shipments before they occur (i.e. waste exporters should notify and receive permission from waste importers before any shipment takes place). For the time being, the Convention does not fully address waste exports intended for recycling, and there is great danger that some exporters may take advantage of such a loophole. However, it delineates what constitutes 'illegal traffic' in hazardous wastes and condemns such traffic as a criminal activity that is subject to international law.

The Basel Convention reaffirmed that the basic principles for the management of wastes must be first, to prevent and reduce the generation of wastes, and secondly, to increase the proportion of wastes that are recycled or reused or treated, so as to reduce their hazardous characteristics. It recommended that the OECD Member countries promote the establishment of appropriate disposal facilities for the management of hazardous wastes at the national level since such action may serve to reduce the need for transfrontier movements of hazardous wastes. The costs of appropriate management and treatment of hazardous wastes must be borne by the generators and other persons who handle or manage the wastes and not by society or the environment. Overall, the Convention addresses the following particularly important general items (Tolba, 1990; Adede, 1993).

- It stresses the principle of sovereign right of every country to ban the import of hazardous wastes.
- It obliges every contracting party to ensure that no shipment of hazardous wastes is allowed to go to a country that has banned the import of such wastes.
- Every country has the obligation to reduce the generation of hazardous wastes to a minimum, and to dispose of such wastes as close as possible to the source of generation. Transboundary movements of hazardous wastes may take place only as an exception, if they present the most environmentally sound solution, and if they are carried out

in accordance with the strict control system provided by the Convention.

- Transboundary movements of hazardous wastes carried out in contravention of the provisions of the Convention are considered to be illegal traffic. It states that illegal traffic in hazardous wastes is criminal, and requires that contracting parties legislate to prevent and punish illegal traffic.
- Provides for bilateral, multilateral or regional agreements to prohibit or regulate movements of wastes.
- Ensures exchange of information and international cooperation regarding transboundary movements of hazardous wastes, including the principle of notification and PIC.
- Requires that the exporting country must be assured that shipments will be disposed of in an environmentally sound manner.
- Recommends that developed/industrialized countries provide technical assistance to developing countries in all aspects of waste management.

Other significant provisions of the Basel treaty indicate that exporters are required to provide detailed information about each waste shipment to the recipient/importing country and to all corridor countries. This information must be sufficient to allow the affected countries to conduct an assessment of the risks involved. Also the agreement requires signatories to immediately inform affected countries of potential risks resulting from accidents during the international shipment of hazardous wastes. The pact further calls for the establishment of a revolving fund (what could be called a 'global superfund') to assist in minimizing damage in emergency situations arising from accidents occurring during transboundary movements of hazardous wastes.

In support of programmes that involve transboundary shipments of wastes, the Basel Convention reaffirmed the fact that efficient and environmentally sound management of hazardous wastes may indeed justify some transfrontier movements of such wastes; however, in all cases, the wastes should be directed to adequate disposal facilities. On the other hand, prior to his retirement from the UNEP at the end of 1992, Dr Mostafa Tolba (Executive Director of the UNEP) had proposed/advocated a complete ban on waste shipments to developing countries and Eastern Europe, citing that the Basel treaty may not be a panacea to this global problem of so-called 'toxic terrorism', in view of the fact that hazardous wastes will likely always follow the path of lower costs and lower environmental standards.

5.3.1.2 The challenges to implementing the Basel Convention

Controlling the transfrontier movements of wastes is an important agenda for the international community. However, there currently is no international machinery to **effectively** police and monitor the hazardous waste trade. Yakowitz (1990) describes in an elaborate manner, important details of the background and extent of problems of transfrontier movements of hazardous wastes; the forces acting to promote transfrontier movements of hazardous wastes; and actions being taken to monitor and control transfrontier movements of wastes, as well as a prognosis of these actions. Indeed there have been considerable differences of opinion with respect to what approaches to adopt for controlling the transfrontier movements of hazardous wastes, especially for shipments from industrialized to developing countries. There are advocates for a total ban of such exports (except where, in some cases, bilateral or multilateral agreements exist between governments with specific terms negotiated and adopted); others suggest that a system for orderly control of transfrontier movements of hazardous wastes is preferable to an outright ban (Yakowitz, 1990).

With all the skirmishes and petty quarrels, suffice it to say that it is difficult to predict when there will be full and effective implementation of this UNEP global system to the satisfaction of the international community. Upon implementation and enforcement, the global Convention has the goal of ensuring the effective and appropriate monitoring and control of transfrontier movements of hazardous wastes internationally. But then, as Yakowitz (1990) indicates, one caveat will be necessary, namely: 'There must be active, strong and effective domestic waste management systems in force.' In the absence of effective domestic control systems (as is typical in most developing countries) to mesh with the international system, adequate protection may not be provided by the Convention, and the goals of the Basel treaty will become elusive. The validity and importance of this observation for the developing world is self-evident.

In any case, despite their general good intentions, international conventions are symptomatic of the cross purposes and varying interests between developed countries and developing nations concerning the hazardous waste trade. There are no guarantees that under these conventions, recipient countries will be fully informed of the risks involved. Disposal procedures in the recipient country are unlikely to be closely scrutinized so that risks to population and the environment can be minimized. As a corollary, the fees paid to developing countries for disposing of these wastes may not fully reflect their associated long-term health and environmental risks. In fact, the Convention does not necessarily ensure that wastes will not be shipped to the most vulnerable cash-starved economy, where low disposal fees are achieved at the expense of increased health risks. Ultimately the responsibility for technology transfer between industry and the waste recipient country could be lost, and exporters may not be assured that their wastes will be handled and disposed of in the most cost-effective manner.

In the very long term, a number of externalities – such as the General Agreement on Tariffs and Trade (GATT) – will probably affect the future agenda for the Basel Convention. For instance, some aspects of the Basel Convention may run foul of GATT's most-favoured-nation (MFN) and national treatment obligations (Wold, 1996). Although the status of hazardous and other wastes as a 'good' or a 'service' is not clear under GATT, and none of the programmes have advanced the notion that hazardous wastes should be handled as a 'commodity' or a 'product', at least not as yet, wastes could eventually be seen as commodities with usable materials for recovery and reuse (Ginn, in Avogadro and Ragaini, 1994; Wold, 1996). At that point in time, a ban on the exports of hazardous wastes will be seen as a non-tariff barrier or an effort on the part of the industrialized nations to limit the access of emerging economies to a new kind of commodity (namely hazardous wastes, especially those with recoverable materials). In fact, if and when waste is seen and accepted as a 'good'/'commodity'/'product', and waste disposal as a 'service', then the Basel Conventions' restraints on exports may raise (among other things) MFN obligation questions (Wold, 1996). It has already been argued that both an importing and an exporting country violate GATT Article XI if they prohibit trade in waste materials because the importing country cannot manage the waste in an environmentally sound manner; that the exporting country also violates its Article I MFN obligation if it permits exports to one Basel Convention party and not another; and that a decision to prohibit trade because the wastes would not be used as raw materials for recycling or recovery industries in the country of import violates GATT Article XI, as well as its MFN obligation, if trade with other countries is permitted (Wold, 1996).

It is apparent that the Basel Convention – and indeed many other multilateral environmental agreements (see, e.g., sections 5.4 and 5.5) – is using a variety of import and export restrictions to protect the environment. Whether such an approach will prove effective should be seen as a major concern, considering the possible conflicts with pre-exiting agreements such as GATT. In fact, by their very nature, these restrictions which specify the types of materials that may be traded are inconsistent with aspects of Article XI of GATT, whereas the restrictions on whom trade activities may be conducted with violate Article I MFN requirements. It must therefore be recognized at this time, that the Basel Convention import–export restraints on hazardous materials/wastes could eventually lead to some kind of trade dispute in the global market-place.

5.4 CONTROL MEASURES BY DEVELOPING COUNTRIES

Several developing countries have not been convinced that the Basel Convention offered protective enough measures for their regions.

Consequently, some have adopted various control measures in relation to the transfrontier shipments of wastes. For instance, the OAU, which currently includes all African countries except Morocco and South Africa, at a summit in Addis Ababa, Ethiopia, in May 1988, adopted a resolution barring any waste trade deals with the industrialized countries. This was followed by meetings that subsequently resulted in the drafting of an 'African Convention on the Transboundary Movement of Hazardous Wastes and Their Disposal'. This African Convention will, among other things, ban all forms of waste shipments to Africa.

At the 11th Summit of the Economic Community of West African States (ECOWAS) held in Lomé, Togo, in June 1988, the waste trade emerged as a major issue of concern. It was recognized that all striving for economic development and progress in the region would be in vain if due consideration were not given to the health of the populations and the environment. ECOWAS leaders at the summit adopted a resolution condemning the toxic waste trade, and agreed to enact national legislations against dumping of foreign wastes in the region.

The climax of all these measures was that, from July 1988, several countries started enacting very strong and strict national policies regarding the waste trade. Furthermore, an OAU ministerial meeting at Bamako, Mali, in January 1991, culminated in the adoption of the 'Bamako Convention on the Ban of the Import into Africa and the Control of Transboundary Movement and Management of Hazardous Wastes within Africa' which, among other things:

- bans the import, for any reason, of hazardous and radioactive wastes;
- requires states that are parties to the Convention to police their territorial seas and waters in order to prevent non-parties from dumping hazardous wastes in the region;
- requires hazardous waste generation audits;
- imposes strict, unlimited, as well as joint and severe liability on hazardous waste generators;
- forbids the import of hazardous substances which have been banned, cancelled or refused registration, or voluntarily withdrawn in the country of manufacture for human health or environmental reasons;
- calls for the issue of the transfer of polluting technologies to Africa to be under careful and systematic scrutiny.

The Bamako Convention, which is open for ratification or accession to all OAU Member States, was drafted with the intention of establishing stricter criteria for the control and management of hazardous wastes than that provided by the Basel Convention.

To further protect West Africa from dumping of toxic wastes, ECOWAS has established a monitoring authority called 'Dumpwatch', to look out for potential or would-be waste dumpers and other violators. Overall, reactions in the African media to this 'toxic terrorism', and 'garbage imperialism' – as some have come to brand the transboundary shipments of wastes from the industrialized worlds into the developing and emerging economies – has been very strong, with most viewing the waste trade/exports as a new form of neo-colonial exploitation (Opsal, 1989).

In other regions, the Caribbean Community (CARICOM) has discussed the issue of international waste trafficking. For instance, at a CARICOM summit in Grenada in July 1989, leaders of 13 Caribbean countries endorsed the Port of Spain Accord, which includes a condemnation of the dumping of hazardous and toxic wastes in the region from areas outside the region (Vallette and Spalding, 1990).

Other regional measures relating to a ban on the toxic waste trade have been initiated/developed in Latin America. For example, the Central American Regional Agreement on the Transboundary Movement of Hazardous Wastes, signed at a summit in Panama in December 1992 by some Central American governments (namely: Costa Rica, El Salvador, Guatemala, Honduras, Nicaragua and Panama, with Belize as an observer country), adopts an approach similar to that of Africa's Bamako Convention that was signed earlier in January 1991, especially insofar as obliging its parties to prohibit the import of hazardous wastes into the Central American region from states that are not parties to the agreement. Indeed a regional import ban forms the centrepiece to the 'Agreement on Transboundary Movement of Hazardous Wastes in the Central American Region'. It bans the importation, transport, ocean dumping and ocean incineration of hazardous wastes in the Central American region. Whereas the law will eliminate almost the entire trade in wastes between Central America and the rest of the world, it will allow certain kinds of wastes to be traded under particular circumstances. The agreement becomes legally binding upon ratification, as set up in the draft legislation.

5.5 RECENT DEVELOPMENTS AND ACTIONS BY THE INTERNATIONAL COMMUNITY

The Basel Convention on the international trade in wastes became law in May 1992 after an initially slow ratification process; this treaty will, at least temporarily, minimize or stop some hazardous wastes from crossing international borders. However, most major waste exporting countries had still not ratified the treaty, thus limiting the effectiveness of the Convention. In any case, the Convention is to be developed further by the adoption of the protocol on liability and compensation for damages resulting from transboundary movements of hazardous wastes and their disposal.

It is noteworthy that the 1992 United Nations Conference on Environment and Development (UNCED) in Rio de Janeiro, Brazil, also addressed some of the issues relating to the international traffic in toxic and dangerous products. In particular, two chapters of Agenda 21 (which is the plan of action for sustainable development into the 21st century and beyond) – namely, Chapters 19 and 22 address these issues. Chapter 19 deals with environmentally sound management of toxic chemicals including the prevention of illegal international traffic in toxic and dangerous products and wastes, whereas Chapter 22 focuses on radioactive wastes.

Recognizing that the arena of international law is a dynamic one, it is not surprising that the First Meeting of the Conference of the Parties to the Basel Convention held in Piriapolis, Uruguay, in 1992, adopted 23 Decisions for the implementation of the Convention. An analysis of these Decisions reveals that the Basel Convention is already developing into a legal international agreement that deals not only with the control of transboundary movements of hazardous wastes, but also that it involves on a larger scale the issue of their environmentally sound disposal as well as technical assistance – achieved mainly through the establishment of training systems, and through the establishment of centres for technology transfers.

The First Meeting of the Conference of the Parties (1992) also adopted: the Lomé IV Convention (1989) amongst the African, Caribbean and Pacific (ACP)/EU; the resolution of the Bamako Convention (1991) on the ban of the import of hazardous wastes into Africa and the control of their transboundary movements within Africa; and the OECD Council Decision (1992) concerning the control of transfrontier movements of wastes destined for recovery operations. As a result, the Conference of the Parties to the Basel Convention requested a total ban on all exports of hazardous wastes from OECD countries to non-OECD countries, because such activities had great potential to be an environmentally unsound way of managing hazardous wastes – as required by the Basel Convention.

The Second Meeting of the Conference of the Parties to the Basel Convention, held in Geneva, Switzerland, in March 1994, adopted 27 Decisions which develop the Convention as well as serve as guidance for its implemention. One of the most important Decisions is that establishing the immediate prohibition of all transboundary movements of hazardous wastes which are destined for final disposal from OECD to non-OCED countries. The transboundary movement of hazardous wastes destined for recycling or recovery operations from OECD to non-OECD states is to be phased out by 31 December 1997, and all such shipments will be prohibited as of that date. At this time, it was requested by the meeting that any non-OECD state that did not have a national hazardous wastes import ban and which allowed the import of hazardous wastes from OECD states for recycling or recovery operation until 31

December 1997, should inform the Secretariat of the Basel Convention that it would allow such imports (from an OECD state of hazardous wastes for recycling or recovery operations) by specifying the following: the categories of hazardous wastes which are acceptable for import; the quantities to be imported; the specific recycling/recovery process to be used; and the final destination/disposal of the residues which are derived from recycling/recovery operations. The question which remains now relates to the means by which the parties may implement this decision proposal in practice. This remained an open question, which was further discussed at the Third Meeting of the Conference of the Parties which took place in 1995.

By 1994 significant progress had been achieved for the implementation of the Basel Convention. Consequently, a relevant Manual was prepared for legal or other persons involved in the generation, export, import and/or disposal of hazardous wastes. A revised Manual was subsequently presented for the consideration of the Third Meeting of the Conference of the Parties in 1995.

In September 1995 delegates attending the Third Conference of the Parties to the Basel Convention in Geneva literally closed the floodgates of toxic waste exports to developing countries. More than 90 parties to the Basel Convention agreed by consensus to amend the Convention to reflect the 'Basel Ban Decision' agreed upon in the previous year. According to that Decision, countries of the OECD could no longer allow the export of hazardous wastes to non-OECD countries for final disposal. And, beginning in 1998, waste exports for recycling will be prohibited as well. The ban amendment will enter into legal force after three-quarters of the parties ratify the amendment into their national law. Meanwhile the EU parliament indicated full support for the implementation; the 15 Member States of the EU were therefore expected to ratify the amendment into EU law.

Up till this point, a series of hazardous waste characteristics have not been adequately defined for the wastes, subject to the Basel Convention; the Secretariat has been continuing its cooperative efforts in the development of such criteria with several organizations, in particular, with the International Maritime Organization (IMO), the United Nations Economic Commission for Europe, the United Nations Industrial Development Organization (UNIDO), the International Programme of Chemical Safety (IPCS), the OECD and the EU. The initial results on progress made were also presented to the Third Meeting of the Conference of the Parties (1995).

Ultimately there is a real need for the establishment of a global waste inventory and database system. The Secretariat of the Basel Convention has been reviewing the relevance of such a system with cooperation from the London Convention (1972), the IMO, the concerned UNEP offices, and the WHO since 1995. The question of whether the international trade in hazardous wastes can be adequately controlled still lingers.

5.5.1 OECD AND EU DEVELOPMENTS, INITIATIVES AND ACTIONS

In 1991 a Decision Recommendation [C(90)178/Final] concerning the reduction of transfrontier movements of wastes was made by the OECD Council. This calls for delineation of such controls as may be appropriate for the transfrontier movements of wastes destined for recovery operations, clarification of the definition of such wastes, and characterization of those wastes which may require different levels of control. The Decision requires that wastes not destined for recovery operations should, to the extent possible, and consistent with environmentally sound and efficient management practices, be disposed of within the Member country in which the waste is generated. The Decision also provides that Member countries should develop a disposal capacity for wastes which currently cannot be managed within their own territory. Alternatively, where this is not possible, countries should enter into bilateral or regional agreements, conducted and approved at governmental level, which provide environmentally sound management for such wastes.

In 1992 the OECD Member countries resolved to create and fully implement an international mechanism to control transfrontier movements of wastes destined for recovery operations within the OECD area. The Decision [C(92)39/Final] contains provisions to allow, what is now referred to as the 'OECD Control System', to be developed. This system identifies wastes destined for recovery operations; classifies the wastes into **Green**, **Amber** and **Red** lists – depending on their overall environmental risk and the management practices utilized; and establishes different levels of control for each category of wastes – ranging from regular commercial control (green) to stricter controls (amber and red). It ensures that sufficient information is provided, in advance, to all countries involved in a transfrontier movement in order that they may decide to allow or prevent the movement of such wastes through their territories.

In order to evaluate the current placement of some wastes on the green, amber or red lists, a review mechanism has been established to make any proposals for revision of the control system (OECD, 1993). As a consequence, the lists of wastes established by the OECD Council Decision C(92)39 in 1992 were revised in 1993. They include the obligation to reduce the generation of hazardous wastes to a minimum, and to ensure that the sovereign right of each state to ban the import of hazardous wastes into its territory is observed. It also prohibits the export and import of hazardous wastes to and from non-parties to the Basel Convention, unless such movement is subject to bilateral, multilateral or regional agreements, or arrangements whose provisions are no less stringent than those of the Basel Convention. It requests that hazardous wastes should be disposed of as close as possible to their source of generation and that the transboundary movement of hazardous wastes could be allowed if it is carried out in accordance with the strict control system, provided by the Convention, which includes prior informed consent by the importing country as well as by the transit country.

In other developments, a year after the Basel Convention, the then 12 Member nations of the EU and some 68 African, Caribbean and Pacific (ACP) countries went a step further by banning exports of toxic and radioactive/nuclear wastes from the 12 EU countries to the 68 developing countries of the ACP league. In turn, the ACP has also agreed **not** to import any wastes from any other non-EU country. Thus, **all of the ACP regions are technically off-limits to international waste trade under the Lomé IV Convention**. The question that remains though, is: If regional facilities are established within the ACP regions, would they be exempt from this condition of 'non-import' from non-EU countries as required under the Lomé IV Convention? Although apparent that inter- and intra-ACP waste trade deals may be allowable, this issue should be clarified before any breach occurs – especially since the establishment of regional hazardous waste management facilities may be an inevitable requirement if the ACP members are to tackle effectively and efficiently the domestic environmental problems engulfing their regions. The use of regional waste facilities could mean possible inter-state movements of some waste types between the ACP members within specified regions. In fact, the provision to prohibit all imports from non-EU states in general apparently further creates a conflict with Africa's Bamako Convention by prohibiting waste movements between African ACP and non-ACP states that are permissible under the Bamako Agreement.

5.5.2 TOTAL BAN ON ALL EXPORTS OF HAZARDOUS WASTES FROM OECD COUNTRIES TO NON-OECD COUNTRIES: IS THIS FOR REAL OR JUST A MIRAGE?

Article 11 of the Basel Convention allows parties to enter into bilateral, multilateral or regional agreements or into arrangements with parties or non-parties regarding the transboundary movements of hazardous wastes or other wastes – provided that such agreements do not detract from the environmentally sound management of hazardous wastes as required by the Basel Convention. Furthermore, such agreements shall stipulate provisions which are no less environmentally sound than those provided by the Convention, in particular by taking the interests of developing countries into account.

The reaction from target countries to the hazardous waste trade – especially those in the developing world – appears to be strong. Unfortunately enforcement is difficult. Politicians and local officials are often corrupt, so that many opportunities exist for circumventing any restrictions. Given the potential for high profits, it is likely that the market for hazardous wastes in developing countries will continue to be profitable and available.

Developing countries are encouraged to participate in the hazardous waste trade by the prospect of much needed foreign exchange. Many of these countries are plagued by lower health standards and low life expectancy. The increased risks associated with the handling and disposal of hazardous wastes may be considered low when compared to everyday risks of living in an impoverished developing country. As a result, tolerance to risk is high, and many opportunities exist for taking advantage of the income disparities between developed and developing countries. The argument can be made that this need not be the case. After all, if the hazardous waste trade is properly managed, such trade programmes can be used to curb poverty, improve quality of life for people in developing countries, and safeguard the environment. All of these objectives can indeed be viewed as compatible terms.

5.6 THE SEARCH FOR A GLOBAL SOLUTION: CAN THE INTERNATIONAL TRADE IN HAZARDOUS WASTES BE CONTROLLED?

Establishing legislation to control international hazardous waste shipments has proved to be far more difficult than was ever envisaged. For example, the First and Second Meetings of the Conference of the Parties to the Basel Convention requested that: the parties promulgate or develop stringent national legislation on the control of transboundary movements of hazardous wastes, taking into account the elements contained in document UNEP/CHWC1/1/7; the parties incorporate in their legal systems, appropriate sanctions or penalties for illegal traffic in hazardous wastes and other wastes; and the regional commissions and protocols take an effective role in the monitoring and prevention of illegal traffic in hazardous wastes and coordinate their efforts and activities to this regard with the Secretariat of the Basel Convention. But even when the Decisions adopted by the Second Meeting of the Conference of the Parties to the Basel Convention (1994) are implemented by Member States, there will still be a considerable shortfall from the objective of full 'cradle-to-grave' control from waste production to final disposal.

A major difficulty with the regulation of international waste shipments relates to the fact that any such transaction will tend to involve a complex network of different countries. For example, EU Member and non-Member countries may become involved in such a transaction by being designated as sources, destinations and/or transit zones, and this greatly increases the complexity of any applicable legislation. Another problem relates to the fact that only classified wastes are controlled by EU states' waste disposal legislation whereas recoverable materials meant for recycling remain uncontrolled. Certain types of recoverable substances go for recovery because the cost is less than the disposal cost of the material as waste. A large number of materials can either be wastes or recyclable, depending on circumstances. The economic value of such materials can be either positive or negative depending mainly upon the current state of the market and price of the recovered materials. Certain solvent recovery companies rely on batch recovery, with some plants processing specific materials at particular times of the year. An increase in waste disposal costs may stimulate recycling (or vice versa). A study in England has shown that even the volume alone of a given material can affect its economic viability for recycling, and thus will affect its waste or goods status. Many types of contaminated solvents fall into this category and a very significant international trade has developed for such materials. In fact, importing wastes under the label of 'recycled goods' could create a loophole for avoiding a country's toxic waste disposal regulations (Laurence, 1988).

In any case, inter-ministerial coordination between respective governments may be necessary in order to prevent and penalize illegal traffic in hazardous wastes. Also, better cooperation with the Customs Co-operation Council Secretariat with a view to ensuring control of the import and export of hazardous wastes subjected to the Basel Convention in the harmonized system may become necessary. Furthermore, it would be useful to receive, collate and disseminate information on the illegal traffic in hazardous wastes and other wastes promptly and systematically, with a view to ensuring that hazardous waste detected and rejected by one country is not dumped in another. It may also prove useful to explore the possibility of cooperation with Interpol and NGOs in cases of illegal traffic. For example, a joint project of the UNEP and the Economic and Social Commission for Asia and the Pacific (ESCAP) on monitoring and assessment of illegal traffic in toxic and dangerous products and wastes has been implemented in that region. Also, it would be advisable to bring to the attention of the parties to the London Dumping Convention, the need for a review of the existing rules, regulations and practices with respect to dumping of hazardous wastes and other wastes at sea.

Finally, it is noteworthy that a major control problem with the cross-boundary shipments of any hazardous wastes may be attributed to the fact that such material is usually considered as having a nil value. There is always a financial incentive for illegally dumping waste loads or shipping them for recycling elsewhere. Thus, maybe, one possible way to reduce the incentive for illegal waste trading is to give the waste a value (Laurence, 1988). This value should be well above the level set for disposal and transportation costs.

Ramifications of a hazardous waste trade 6

The goal of a sustainable and healthy economic growth inherently requires careful environmental consideration. But, the financial crisis in most developing countries in particular – due especially to falling prices and earnings from commodity exports – has, for instance, sharply increased their debt-repayment burdens. By virtue of the fact that such countries are heavily in debt, it becomes difficult to justify any spending of scarce public funds on environmental initiatives. Thus, in most developing countries, it takes an extra strong national leadership, willing to gamble both politically and economically, to appreciate and make policy decisions on environmental issues.

As a result of the numerous financial pressures, some developing and newly industrializing countries have even succumbed to pressures to accept wastes for foreign exchange. Where such actions have taken place, the activities have been carried out without adequate evaluation of the potential long-term impact on the nation's population and environment. As noted by the UNDP (1990), however, '... it is short-sighted to balance budgets by unbalancing the lives of people ... '. In fact, since the lives of people are intricately linked to a sustainable environment, any environmental deterioration caused by economic activities will ultimately impact the lives of populations living in any given region. A country may not, therefore, pursue an economic development activity that will be detrimental to the very population that such development activity is meant to serve. On the other hand, as Robert S. McNamara (former President of the World Bank) noted in his 8 June, 1972 policy address to the United Nations Conference on the Human Environment in Stockholm, Sweden, '... valid environmental considerations need not deny the advance in economic development the less-privileged countries so gravely require'. This is because, much as the poor nations have no desire to see their own environment contaminated or wantonly abused, neither do they have the desire to remain caught in the permanent 'contamination of poverty'.

In any case, there are several detrimental health, environmental and socio-economic impacts that can be attributed to inadequate and inefficient management of toxic wastes in general. In particular, illegally dumped and/or inappropriately disposed of hazardous wastes may enter the environment and threaten human and ecological systems at different levels. The risks to human health include, but are not limited to, such prognoses as cancers, birth defects, nervous system problems, reproductive system damage, cell mutations, etc. As Schwartz and Pratt (1990) point out, the psychological, economic and social consequences of these illnesses – including birth defects, and even deaths – can be devastating to individuals and families directly involved, and costly to society at large. It is understandable, therefore, that institutions like the World Bank, the OAU and the ECOWAS should take a firm stand against the waste trade, since inadequate and inefficient hazardous waste management schemes can add considerably to the problems that face developing countries – although schemes that are effectually managed can be only beneficial to such countries.

6.1 THE NATURE OF HAZARDOUS WASTES AND WASTE CLASSIFICATION

Wastes from industrial processes typically include a wide range of materials which may have varied chemical compositions – such as different proportions of organic and inorganic compounds. In a broad sense, a hazardous material is one which is capable of producing some adverse effects and/or reactions in potential biological receptors; toxic substances generally present unreasonable risk of harm to human health and/or the environment, and need to be regulated. Specifically, hazardous wastes include those by-products with the potential to cause detrimental effects on human health and/or the natural environment; such wastes may be toxic, bio-accumulative, persistent (i.e. non-degradable), radioactive, carcinogenic (i.e. cancer-causing), mutagenic (i.e. causing gene alterations) and/or teratogenic (i.e. capable of damaging a developing foetus). Their general heterogeneity makes their disposal all the more difficult.

Major categories of industrial wastes which are considered hazardous include solvents, waste paint, waste containing heavy metals, acids and oily wastes; wastes from mining activities which may occur as mine tailings that are contaminated by metals and chemicals; and large amounts of ash that are often the product of energy generation and/or incineration processes. The degree of hazard posed by the wastes may be dependent on several factors (Box 6.1).

The identification and categorization of potentially hazardous waste streams are important in the investigation of 'candidate' wastes for a haz-

Box 6.1
Important factors affecting the degree of hazard posed by wastes

- Composition, quantities and physical form.
- Reactivity (including ignitability, explosivity, corrosivity and leachability).
- Biological and ecological effects (including human toxicity, ecotoxicity and radioactivity).
- Mobility (i.e. transport in various environmental media).
- Persistence (including fate in environment, and detoxification potential).
- Indirect health effects (from pathogens, vectors, etc.).
- Local conditions (e.g. temperature, soil type, groundwater table conditions, humidity and light).

ardous waste trade (i.e. those wastes generated in such quantities or forms to prompt possible export) and in assessing effective corrective actions. Also the emergence of international agreements on controlling the transfrontier movement of hazardous wastes prompted national governments to recognize the need for a common list of wastes, specifying the characteristics which make them hazardous. For example, a Core List of 17 generic waste types and 27 constituents was adopted by the OECD countries, and the 15 categories of waste to be controlled by the Basel Convention on the Control of Transboundary Movement of Hazardous Wastes and their Disposal correspond, to a large extent, to this OECD Core List.

The 'candidate' wastes for waste trade schemes generally originate from several industries; the industries generating 'exportable' wastes in the industrialized nations include those previously identified in Table 3.2 (section 3.1). The waste types likely to be candidates for export will normally be extremely hazardous, and are banned from legal sea dumping. In order to develop adequate waste management strategies and waste trade policies, the 'candidate' wastes must be appropriately categorized. A typical categorization will comprise putting the wastes into groupings (e.g. high-, intermediate- and low-risk waste classes, conceptually represented by Figure 6.1) based on their potential health and environmental risks and on their physico-chemical properties. Wastes may also be classified/categorized according to whether they are suitable for a proposed disposal outlet in a country of import – such as using an incinerator to handle organics versus chemical fixation treatment and/or encapsulation for heavy metals.

In any case, the high-risk wastes will prompt the most concern about any waste trade, since they typically are known to contain significant

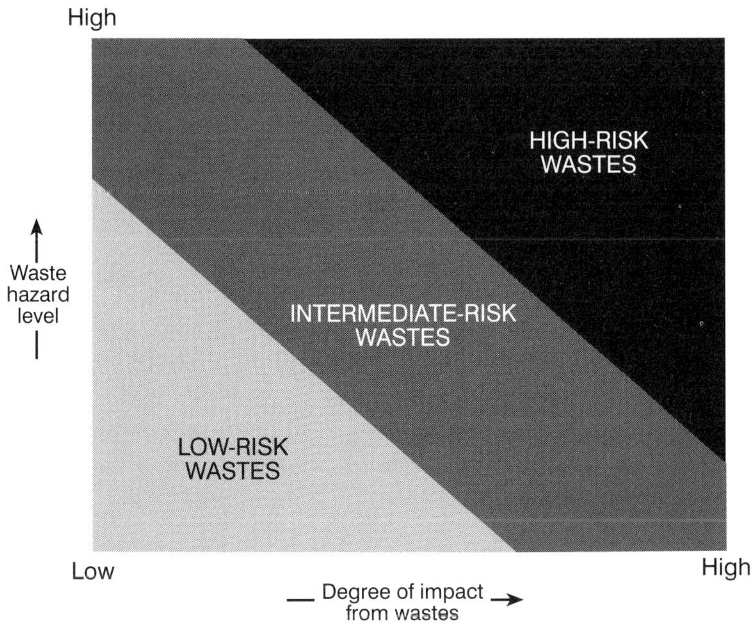

Fig. 6.1 A conceptual representation for hazardous waste classification systems.

concentrations of highly toxic, mobile, persistent and/or bio-accumulative constituents; the low-risk wastes typically include primarily high-volume/low-hazard wastes and some putrescible wastes, for which the cut-off between a 'hazardous' and 'non-hazardous' waste is least clear-cut (Batstone *et al.*, 1989). Examples of high-risk wastes are chlorinated solvent wastes from metal degreasing, cyanide wastes, dioxin-based wastes and PCB wastes; intermediate-risk wastes will typically include metal hydroxide sludges for which the toxic metals are in relatively insoluble physical form with low mobility.

Whatever strategy or scheme is used, it is important to recognize that there are varying degrees of hazards associated with different waste streams, and there are good economic advantages for ranking wastes according to the level of hazard they present. This will facilitate efficient waste management programmes recommended for implementation by a given locality. Also, in a proposed waste trade programme, it should be recognized that if a given transaction falls in the low-risk category for instance, then a different negotiating framework may have to be adopted – compared to one that potentially belongs to the high-risk group.

In general, the classification requirements should also include identifying the compatibilities and/or incompatibilities of the chemical constituents of the 'candidate' wastes to allow effective management. A matrix of waste classes that identifies compatible and incompatible wastes should be developed during the waste categorization process. Potentially incompatible wastes should be identified, together with the potential consequences of such mixtures. In addition, compatible and even 'friendly' wastes should be identified. This will all add up to help in the overall waste management programmes and also waste trade decisions for a given country or region.

6.2 HAZARDOUS WASTE MANAGEMENT RISK ISSUES

Risk – generally interpreted as the probability of a harmful event to humans or to the environment that is caused by a chemical, physical or biological agent – can be described by the following conceptual relationship:

$$\text{Risk} = [f(I) \times f(P)] - f(D),$$

where f(I) represents an 'intrinsic risk' factor that is a function of the characteristic nature of the agent or the dangerous properties of the substance; f(P) is a 'presence' factor that is a function of the quantity of the substance released in the environment and of all the accumulation and removal methods related to the chemical and physical parameters of the product, as well as to the site-specific parameters typical of the environment; and f(D) represents a 'defence' factor that is a function of what society can do in terms of both protection and prevention to reduce to a minimum, the harmful effects of the problem materials. A very important factor in this equation is f(D), which includes both the ordinary defence mechanisms for hazard abatement and the legislative measures which define such factors as the transportation conditions that will help minimize transport-related risks. As a consequence, one should have to take into account during the evaluation of waste management programmes, the present legislative measures in order to assure the safety of the hazardous waste management at all stages of the waste material's life cycle.

6.2.1 RISKS DUE TO 'STATUS QUO' WASTE MANAGEMENT PRACTICES

Any chemical can cause severe health impairment or even death if taken in sufficiently large amounts. On the other hand, there are chemicals which, even in small doses, can cause adverse health impacts. Some of these are believed to be present in a number of environmental matrices and settings in several developing and newly industrializing countries, and are likely to have been discharged in waste streams which emanated from local industries and their operations, or in wastes from other activities.

In fact, in a number of countries with emerging economies, industrial effluents from factories and mills are believed to have discharged or be discharging unmonitored levels of pollutants into rivers; pollutants which could kill fish, could also threaten the health of those consuming seafood from impacted riverine, or could have a devastating impact on potable water supply sources, etc. Populations may also be exposed to several toxic chemicals as a result of indiscriminate dumping of hazardous wastes on land, and general waste management (or rather waste mismanagement) practices. Such wastes may give rise to continuing contamination of several environmental compartments, and consequently pose unmonitored risks of potential concern in some of these regions. This means that, even for the developing countries that may not have been targets for past waste trade activities, there is still concern about possible environmental hazards due to local waste management practices.

Then of course is the added risk from any waste materials imported into a given region or country. Of critical concern is the cross-boundary movements from the industrialized world to the emerging economies. It is believed that much of the wastes shipped to most developing countries in particular arrive without primary treatment to reduce the degree of hazard posed by such wastes. Historically, storage of the imported wastes takes place in an environmentally unacceptable manner, leading to potential risks to the public. Furthermore, waste disposal methods used by most developing countries are significantly inadequate and may present unacceptable risks to populations and ecological systems in these regions. The risks from waste treatment, storage and disposal methods may be due to the release of chemical constituents into air, surface water, groundwater, soils, sediments and biota. Consequently, general inadequacies in overall waste management practices can create potentially hazardous situations and cause great concern to any nation or region.

6.2.2 TRANSPORTATION RISK ISSUES

An additional issue regarding waste management practices that could have far-reaching implications on waste trade schemes relates to transportation risks associated with the movement of the wastes from the point of generation to the TSDFs. Transport risk is indeed of major concern, particularly with regard to long-hauling of wastes – i.e. transport risks tend to increase with distance. In addition, there are several potential environmental consequences of the transport process itself (for example those associated with fuel usage, air pollution problems, global warming issues, etc.). In fact, transport problems can be a key issue for hazardous waste export, not only from the point of view of economics and risk assessment, but also from the social and psychological perspectives as well. The 'tell-all-tale' can be determined from records of spills of virgin chemicals, oils, etc. during transportation, and indeed other transportation accidents recorded in the literature elsewhere (e.g. Saccomanno *et al.*, 1989; Davidson, 1990).

The risks of transporting hazardous wastes may be defined in terms of the accident probability, the release probabilities in an accident situation, the hazard classes for different damage scenarios, and the expected consequences on populations and environment within the accident/impacted corridor. Different levels of risks may be associated with different shipments that depend on the waste properties/categories, the spill/release scenarios and the overall transportation environment. Considering the relatively poor transportation network in most developing countries, serious consideration would have to be given to transportation routes and associated potential risks and costs when planning for waste management facilities in these regions. In particular, decisions to operate regional hazardous waste facilities should seriously consider the implications of transporting wastes over long distances.

In fact, for countries involved in the international waste trade, transportation risks associated with transporting wastes from the point of entry to potential TSDFs is an important issue which cannot be played down. There are several examples of transport and distribution accidents which should be taken into account in the evaluation of waste trade hazards. The transport of hazardous chemicals by road, rail and barge is a major risk to environmental and human health, the importance of which continues to grow as both the amount of traffic and the quantities of wastes requiring transportation increase. Typically, releases of toxic, flammable and explosive substances may occur following collisions and accidents resulting from mechanical and operational failures. Nearly 85% of accidents at sea involve tankers and there is a growing tendency for such accidents to happen near coastlines – hence the high potential for ecological damage from this source. There are also other aspects of marine spills, namely the unknown impact of spilled toxic chemicals.

6.2.3 AN 'ACCEPTABLE' RISK CONCEPT

An important issue in risk assessment is the risk acceptability level, i.e. what level of risk society can allow for a specified hazard situation, also recognizing that the desirable is not always attainable. With maintenance of public health and safety being crucial, it should be realized that financial constraints may not by themselves be justifiable enough reason for setting acceptable levels on the higher side of the risk spectrum.

An important concept in risk management – especially in relation to human health risks – is that there are levels of risk that are so great that they must not be allowed to occur at all cost, and there are other risk levels that are so low that they are not worth bothering with even at insignificant costs; these are known, respectively, as *de manifestis* and *de minimis* levels (Travis *et al.*, 1987; Whipple, 1987; Kocher and Hoffman, 1991; Suter, 1993). Risk levels between these bands are typically balanced against costs, technical feasibility of mitigative actions, and other socio-economic, political and legal considerations, in order to determine their acceptability or tolerability.

It is noteworthy that the concept of *de manifestis* risk is not controversial because, after all, some hazard effects are clearly unacceptable, whereas the *de minimis* risk concept tends to be controversial in view of the implicit idea that some exposures to and effects of pollutants or hazards are acceptable (Suter, 1993). In any case, it is still desirable to use these types of criteria to eliminate obviously trivial risks from further risk management actions – considering that society cannot completely eliminate or prevent all human and environmental health effects associated with environmental contamination problems. In fact, every social system has a target risk level – whether it is explicitly indicated or not – that represents tolerable danger limits which it is prepared to accept in consequence of potential benefits that could accrue from a given activity; this may be represented by the *de minimis* or 'acceptable' risk level.

Risk is *de minimis* if the incremental risk produced by an activity is so small that there is no incentive to modify the activity (Fischhoff *et al.*, 1981; Covello *et al.*, 1986; Whipple, 1987; Cohrssen and Covello, 1989). These are levels judged to be too insignificant to be of social concern or to justify use of risk management resources to control them, in view of other beneficial uses for the limited resources available in practice. Simply stated, the *de minimis* principle assumes that extremely low risks are trivial and need not be controlled. A *de minimis* risk level would therefore represent a cut-off, below which a regulatory agency could simply ignore alleged problems or hazards. Thus, in the process of establishing risk levels, it is possible to use *de minimis* levels below which one need not be concerned (Rowe, 1983).

The concept of *de minimis* or acceptable risk is essentially a threshold concept, in that it postulates a threshold of concern below which there would be indifference to changes in the level of risk. In fact, considerable controversy exists concerning the concept of 'acceptable' risk in the risk and decision analysis literature. In practice, acceptable risk is the risk associated with the most acceptable decision – rather than being acceptable in an absolute sense. It has been pointed out (Massmann and Freeze, 1987) that acceptable risk is decided in the political arena and that 'acceptable' risk really means 'politically acceptable' risk. Current regulatory requirements are particularly important considerations in establishing acceptable risk levels.

In general, the selection of a *de minimis* risk level is contingent upon the nature of the risks, the stakeholders involved and a host of other contextual variables (such as being compared against other risks). This means that *de minimis* levels will be fuzzy (in that they can never be precisely specified), and relative (in that they will depend on the special circumstances). Also, establishing a *de minimis* risk level is often extremely difficult because people perceive risks differently. Moreover, the cumulative burden of risks could make a currently insignificant risk become significant in the future. Consequently, stricter *de minimis* standards will usually become necessary in dealing with newly introduced risks affecting the same population groups.

There are several approaches to deriving such *de minimis* levels, but that which is selected should be justifiable based on the expected socio-economic, environmental and health impacts. A common approach in placing risks in perspective is to list many risks (which are considered similar in nature) along with some quantitative measures of the degree of risk. Typically, risks below the level of a one in a million (i.e 10^{-6}) risk of premature death will often be considered insignificant or *de minimis* by regulatory agencies, since this compares favourably with risk levels from 'normal' human activities (e.g. 10^{-3} for smoking a packet of cigarettes per day, or rock climbing; 10^{-4} for heavy drinking, accidents in the home, driving motor cars, farming; 10^{-5} for lorry driving, household fires, skiing, use of contraceptive pills; 10^{-6} for diagnostic X-rays, fishing; and 10^{-7} for drinking about 10 litres of diet beverage containing saccharin). In considering a *de minimis* risk level, the possibility of multiple *de minimis* exposures with consequential large aggregate risk should not be overlooked. Whipple (in Paustenbach, 1988) suggests the use of a *de minimis* probability idea for its application and to help develop a workable *de minimis* policy.

De minimis is a lower band on the range of acceptable risk for a given activity. When properly utilized, a *de minimis* risk concept can help to concentrate attention with respect to risks in a socially beneficial way. It may define the threshold for regulatory involvement. In determining a target risk level associated with a possible waste trade activity, a more practical and realistic acceptable risk level ought to be specified. It is only after deciding on an acceptable risk level that a waste trade deal can be effectively evaluated for any potential benefits.

6.3 HEALTH AND ENVIRONMENTAL IMPLICATIONS OF WASTE TRADE PROGRAMMES

If dumped indiscriminately in any environmental media, hazardous wastes may have both short-term and long-term effects on human receptors and ecological systems. Wastes and their disposal can indeed be the cause of several environmental pollution problems. Unfortunate lessons from several countries, such as the Love Canal incident in New York, USA (see section 3.4), clearly demonstrate the dangers arising from unsafe disposal and/or management of hazardous wastes. Analogous incidents are recorded in other US locations, in Europe, in Japan and other locations in Asia (Asante-Duah, 1993).

Exposure to constituents of the candidate wastes in the global waste trade markets may produce adverse effects in both humans and ecosystems. Improper treatment, storage and disposal of the wastes can result in release of contaminants, possible exposure, and potential adverse health and environmental impacts. The potential for adverse health effects on populations from contact with hazardous wastes may involve any organ system, and will depend on the specific chemicals contacted, the extent of exposure (i.e. dose or intake), the characteristics of the exposed individual (e.g. age, gender, body weight, psychological status, genetic make-up, immunological status), the metabolism of the chemicals involved, weather conditions, and the presence or absence of confounding variables such as other diseases (Grisham, 1986). In general, several health effects of primary concern may result from populations exposed to hazardous chemicals due to inefficient management of locally produced or imported wastes (Box 6.2).

Table 6.1 lists typical symptoms, health effects and other biological responses that may be produced due to the presence of specific toxic chemicals in the environment. Indeed exposures to chemicals escaping into the environment can lead to a reduction of life-expectancy and possibly a period of reduced quality of life (caused by anxiety from exposures, diseases, etc.). An uncontrolled waste trade can therefore be perceived as a potential source of several health and environmental problems and of no benefit to the developing countries. Indeed the potential health risks, environmental impacts or other long-term implications

Box 6.2
Typical/major health effects associated with the presence of hazardous materials in the environment

- Carcinogenesis (i.e. causing cancers).
- Genetic defects, including mutagenesis (i.e. causing alterations in genes which are transmitted from generation to generation, or, causing heritable genetic damage).
- Reproductive abnormalities, including teratogenesis (i.e. causing damage to developing foetus).
- Alterations of immunobiological homeostasis.
- Central nervous system (CNS) disorders.
- Congenital anomalies.

Table 6.1 Some typical toxic manifestations and effects caused by some environmental chemicals.[a]

Chemical	Typical health effects and toxic manifestations
Arsenic and compounds	Acute hepatocellular injury, anaemia, angiosarcoma, cirrhosis, developmental disabilities, embryotoxicity, heart disease, hyperpigmentation, peripheral neuropathies
Asbestos	Asbestosis (scarring of lung tissue), fibrosis (lung and respiratory tract), emphysema, irritations, pneumonia, pneumoconioses
Benzene	Aplastic anaemia, CNS depression, embryotoxicity, leukaemia and lymphoma, skin irritant
Beryllium	Granuloma (lungs and respiratory tract)
Cadmium	Developmental disabilities, kidney damage, neoplasia (lung and respiratory tract), neo-natal death/foetal death, pulmonary edema; bio-accumulates in aquatic organisms
Chromium and compounds	Asthma, cholestasis (of liver), neoplasia (lung and respiratory tract), skin irritant
Copper	Gastro-intestinal irritant, liver damage; toxic to fish
Cyanide	Asthma, asphyxiation, hypersensitivity, pneumonitis, skin irritant; toxicity to fish
Dichlorodiphenyl trichloroethane (DDT)	Axtaxic gait, convulsions, human infertility/reproductive effects, kidney damage, neurotoxin, peripheral neuropathies, tremors; bio-accumulation in aquatic organisms
Dieldrin	Convulsions, kidney damage, tremors; bio-accumulation in aquatic organisms
Dioxins	Hepatitis, neoplasia, spontaneous abortion/foetal death; bio-accumulative
Formaldehyde	Allergic reactions; gastro-intestinal upsets; tissue irritation
Lead and compounds	Anaemia, bone marrow depression, CNS symptoms, convulsions, embryotoxicity, neoplasia, neuropathies, kidney damage, seizures; bio-magnifies in food chain
Lindane	Convulsions, coma and death, disorientation, headache, nausea and vomiting, neurotoxin, paresthesias

Lithium	Gastro-enteritis, hyperpyrexia, nephrogenic diabetes, Parkinson's disease
Manganese	Bronchitis, cirrhosis (liver), influenza (metal-fume fever), pneumonia
Mercury and compounds	Ataxic gait, contact allergen, CNS symptoms; developmental disabilities, neurasthenia, kidney and liver damage, minamata disease; bio-magnification of methyl-mercury
Methylene chloride	Anaesthesia, respiratory distress, death
Napthalene	Anaemia
Nickel and compounds	Asthma, CNS effects, gastro-intestinal effects, headache, neoplasia (lung and respiratory tract)
Nitrate	Methemoglobinemia (in infants)
Organochlorine pesticides	Hepatic necrosis, hypertrophy of endoplasmic reticulum, mild fatty metamorphosis
Pentachlorophenol (PCF)	Malignant hyperthermia
Phenol	Asthma, skin irritant
Polychlorinated biphenyls (PCBs)	Embryotoxicity/infertility/foetal death, dermatoses, hepatic necrosis, hepatitis, immune suppression; toxicity to aquatic organisms
Silver	Blindness, skin lesions, pneumonoconiosis
Toluene	Acute renal failure, ataxic gait, CNS depression, memory impairment
Trichloroethylene	CNS depression, deafness, liver damage, paralysis, respiratory and cardiac arrest, visual effects
Vinyl chloride	Leukaemia and lymphoma, neoplasia, spontaneous abortion/foetal death, tumors, death
Xylene	CNS depression, memory impairment
Zinc	Corneal ulceration, oesophagus damage, pulmonary edema

[a](Sources: Various, including: Rowland and Cooper, 1983; Blumenthal, 1985; Grisham, 1986; Lave and Upton, 1987.)

arising from a waste trade activity may exceed the short-term economic gains by several orders of magnitude. Furthermore, improper waste management practices may reduce the future development potential of the country accepting wastes in a trade deal. Potential health and ecological problems are averted, however, by carefully implementing substantive management programmes for waste disposal.

6.4 SOCIO-ECONOMIC INDICATORS OF WASTE TRADE IMPACTS

Several studies have arrived at the broad consensus pointing to socio-economic status or economic prosperity (measured in terms of income levels, GDP, etc.) as the key determinant of overall health status and social well-being of a society (Lind *et al.*, 1990; Nathwani *et al.*, 1990b). Indeed, it has been argued that increasing the incomes of countries has the potential to increase safety far more than all risk reduction policies. Developing countries may therefore hope to attain higher standards of living by collaborating in the waste trade.

Assuming that funds are allocated in a socially responsible manner, changes in the GDP through engagement in the hazardous waste trade can be used to offset increased risks. But in general, all risk mitigation undertaken during a hazardous waste management or trade activity should help maximize the total net benefits in terms of quality-adjusted life-expectancy (QALE). However, maximization of the group benefits should not be at the expense of any identifiable individual or group (at least not without such individual or group being adequately compensated). Nathwani *et al.*(1990b) have developed a compound indicator of QALE called the life product (LP), which is a function of two simple aggregated social indicators – the real GDP per person per year and the life-expectancy (LE) at birth. The proposed relationship is given by:

$$LP = GDP^{\alpha}LE,$$

where α is a parameter that is a function of time. Nathwani *et al.* (1990b) suggest a value of $\alpha = 1/6$ as appropriate for many contemporary developed societies. This value is said *not* to be so sensitive to the degree of development of a society, and so $\alpha = 1/6$ is also appropriate for use for the developing world, though it may be slightly higher in this case. The LP may serve as a measure of benefits to the public. The value of the index α is a reflection of the value placed on a reduction of mortality in terms of economic expenditures.

Ultimately, a family of LP indicators can be derived from three basic well-known social indicators – the population, the real GNP or GDP, and the LE at birth; which can serve as a quality-adjusted measure of LE (Nathwani *et al.*, 1990b). This means that a LP may be used as an objective function in making decisions on a given activity or policy. And in applying this concept to the waste trade, the anticipated increase in real GNP of a country from participating in the waste trade would be the dominant variable in trying to establish a new LP, and in making a final policy decision.

Another possible socio-economic indicator that may be applicable to this type of policy analysis is the human development index (HDI), which is described in the UNDP's *Human Development Report 1990* (UNDP, 1990). The basic/primary variables in the formulation of the HDI are LE, literacy and real GDP per capita; these together with several other secondary variables and constraints help define the HDI as a socio-economic parameter for policy decisions.

As UNDP (1990) notes, poverty is one of the greatest threats to the environment; the concomitant environmental damage then reinforces poverty, and the cycle of economic and environmental degradation continues. Indeed many choices that degrade the environment are made by the developing countries because of the urgency to fulfil immediate needs and for survival. Any plans of action for global environmental protection and/or improvement must therefore include programmes to reduce poverty in the developing world (if this is to become a successful programme).

In fact, most developing economies face severe financial constraints – as reflected by some basic socio-economic indicators for selected countries in Africa, the Americas and Asia (Table 6.2). Generally, poor economic status places individuals at greater risks; for example, inadequate health care, poor nutrition and a potentially degraded environment for some of the impoverished nations in the developing world may be the reason for an average LE at birth that is approximately 20 to 30 years lower than that experienced in most industrialized countries. On the other hand, it is apparent that an improvement in the economic status (measured by the GNP per capita) has the potential to increase the LE (or reduce the mortality risk) (Figure 6.2). Consequently, for most developing countries, the importing of wastes could be seen – rightly or wrongly – to represent an important supplemental source of income needed to boost their economy or raise their GNP. Of course, this view may be held by some countries without recognition of the potential dangers and long-term negative effects. The important question, however, is whether the waste trade will act to increase existing imbalance or can be used to solve some of the urgent economic problems and human needs, without introducing excessive and 'unacceptable' new risks.

In any case, a cultural set-up in which issues relating to public safety are completely ignored is apparently a matter of the past. All nations are becoming increasingly aware and sensitive to the dangers of inadequacies in hazardous waste management practices, in particular with respect to disposal practices that involve the transfrontier movements of wastes.

Table 6.2 Some basic social and economic indicators for selected developing countries, 1987.[e]

Country	Area ('000 sq km)	Population (thousands)	Population density (per sq km)	GDP ($m.)	Real GDP per capita [actual] (purcasing-power parity, $)	Industrial sector % to GDP (approx.)	GNP per capita (US$)	Health expenditure (approx., % GNP)[a]	Total external debt[b] (US$m.)	Life-expectancy at birth[c] (years)
Angola	1 247	9 500	8	n/a[d]	1 000	n/a[d]	470	1.0	n/a[d]	45
Bangladesh	144	110 000	764	17 600	883	23	160	0.6	7 868	52
Benin	113	4 300	38	1 570	665	18	310	0.8	890	50
Burkina Faso	274	8 300	30	1 650	500	40	190	0.9	665	48
Brazil	8 512	144 000	17	299 200	4 310	68	2 020	1.3	110 675	65
Ethiopia	1 222	45 000	37	4 800	454	25	130	1.0	2 139	42
Ghana	239	13 600	57	5 080	481	26	390	0.3	2 385	55
Guinea	246	6 500	26	2 166	500	37	320	1.0	1 516	43
India	3 288	820 000	249	220 800	1 053	48	300	0.9	41 088	59
Indonesia	1 919	175 000	91	69 700	1660	46	450	0.7	42 090	57
Ivory Coast	322	11 100	34	7 650	1 123	41	740	1.1	10 865	53
Kenya	583	23 000	39	6 900	794	32	330	1.1	4 504	59
Liberia	111	2 300	21	990	696	33	450	1.5	1 303	55
Mexico	1 973	85 000	43	141 900	4 624	65	1 830	1.7	101 722	69
Nigeria	924	106 600	115	24 390	668	51	370	0.4	21 876	51
Senegal	196	7 000	36	4 720	1 060	44	520	1.1	2 990	48
Sierra Leone	72	3 800	53	900	480	23	300	0.7	590	42
Zimbabwe	391	9 200	24	5 200	1 184	76	580	2.3	2 481	59

[a]Health expenditure as % GNP for most developed countries is approximately 5.0 to 8.0%.
[b]1986.
[c]Average life-expectancy at birth for developed countries is estimated at approximately 75 years; common range is 70–78 years.
[d]Not available.
[e](Sources: Various, including: World Bank, 1989b; UNDP, 1990; Lind, Nathwani and Siddall, 1991.)

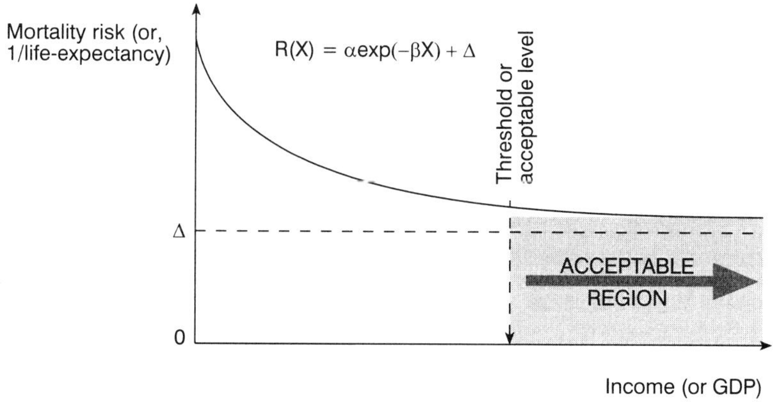

Fig. 6.2 Illustration of the dependence of life-expectancy (or mortality risk) on income levels.

The idea of 'what you don't know doesn't hurt' is no longer a solution to the siting of hazardous waste TSDFs, since even the least-educated populations have some conscious awareness of the potential dangers associated with hazardous wastes. Consequently, in addition to managing the actual risks associated with such facilities, professionals must also deal with the psycho-social effects that the mere presence of toxic materials may have on communities (Peck, 1989).

Naturally, economic growth or development directed to reduce poverty and improve living standards in any country should *not* be achieved at the expense of environmental sustainability for the country or region. Again, the assumption made here is that an increase in GNP through participation in a hazardous waste trade would be used to offset any incremental risks due to the activity, and to even help reduce the 'status quo' risks in society. This is not to suggest that an increase in national income as measured by the GNP means economic growth and/or economic success, since GNP is only one of several factors affecting the level of welfare in a nation; other important factors such as income distribution will have substantial influence on the general welfare status.

Whatever the reason for adopting such a policy, if a waste trade activity or a waste management practice is suspected to present escalated risks, then immediate action should be taken to abate the situation. As Nathwani *et al.* (1990a) indicate, risk reduction efforts in the control of technology are primarily driven by considerations of potential adverse health impacts, so that, a quantitative indicator of safety based on mortality statistics is a reasonable measure of the effectiveness of such measures. Accordingly LE and LP can be used as surrogate measures for safety, so that detriments or risks associated with a waste trade activity or a waste management programme can be assessed in terms of potential loss of LE (LLE) for risk-costs. The benefits are assessed in terms of potential gain in LE (GLE). The use of these measures results in the use of commensurate units or common currency for risks, costs and benefits.

Indeed LE and LP are reasonable indicators of the overall health status and overall quality of life, respectively, of a population. This means that if an activity (e.g. engagement in a hazardous waste trade programme) can result in an improvement in the quality of health care available, improved living standards, and better quality of services received by the population, **without imposing excessive new risks**, then it might be argued that it is a worthwhile activity to consider in an overall economic development programme. A measure of the quality of life can be assumed to be the level of real GDP per person, or indeed the real GNP, if one assumes that people and society will allocate income-earning capacity to enhance what they view as quality of life and values.

6.5 COMPENSATIONS FROM A HAZARDOUS WASTE TRADE

Even though all nations need to integrate environmental issues into their economic development programmes, apparent short-term incentives often prevent some countries from confronting, and dealing with, longer-term environmental problems associated with economic development policies. One area of concern involves the international trade in hazardous wastes. In fact, the quest by developing countries to attain higher living standards, and reach their economic development objectives may involve collaboration in the waste trade that has plagued some of them. The main ingredients of an effective waste trade programme for the waste recipient country are equitable compensation, minimum risks and responsible management/leadership.

The question of responsible leadership is not easily assessed. After all, there can be no guarantee that governments will use funds earned from this type of programme to improve the well-being of their citizens. But, responsible management can be assisted by operating in an open forum, such that all the facts of a given trade agreement are fully divulged. Part of the responsibility for implementing such a regulatory scheme may have to rest on the industrialized world, and on existing international bodies such as the UN, thus allowing for international censure.

In any case, it is apparent that most of the developing and less industrialized countries that have been receiving wastes from industrialized countries are generally not adequately compensated. Hazardous waste trade that is carried out without appropriate compensation is a poor economic development policy. In the long term, therefore, the economic arguments for participating in the waste trade may be defeated. Each programme needs to be judged on its own merits, based on the overall net gains (with respect to risks, costs and benefits) to society resulting from the specific trade activity.

Appropriate compensation for waste imports must recognize the full costs of risks, remedial actions and other direct and indirect costs borne by the waste recipient country, as well as make insurance an integral part of the total compensation package. To help establish a framework for determining equitable compensation to waste recipient countries, consider two illustrative case regions. Region A is an industrialized economy and the waste generator that is seeking to export wastes elsewhere; Region B is a developing country potentially targeted to receive wastes from the industrialized world. It is assumed that wastes exported from Region A to Region B will be treated, stored and disposed of in a safe and environmentally sound manner – this being a precondition for the waste trade agreement. Consequently, significant capital will be spent to achieve an acceptable level of risk at the TSDFs that is appropriate for the site-specific characteristics and conditions at Region B (the waste recipient country). Figures 6.3a and 6.3b demonstrate the conceptual ideas and guidelines that will facilitate the development of a framework and quantitative relationships or formulas for determining equitable compensation in any international trade in hazardous wastes. Figure 6.3a illustrates a case where the same levels of risk tolerance or acceptable risk are assumed for both Regions; Figure 6.3b illustrates a case whereby a higher risk tolerance is assumed for Region B – a reasonable and fair case for allowing double standards – in view of the generally higher 'status quo' risk levels in these regions.

To allow for equity in any perceived waste trade programme, mitigation, compensation and insurance are major incentives that should be used by negotiators in all transactions. Whereas mitigation incentives focus on control of physical or perceived risks, compensation measures focus on equity and fairness in distribution of accrued benefits, insurance gives further guarantees pertaining to long-term liabilities. Although these are all considered complementary, mitigative issues should precede negotiations for compensation and insurance. Compensation may indeed take several forms. O'Hare *et al.* (1983) suggest the following.

- Monetary, that may help reduce tax burdens etc.
- Conditional compensation, that guarantees/insures property values and also offers insurance protection.
- In-kind compensation, that provides alternatives and/or substitutes for previous land-uses.
- Protection of health and safety, including emergency response.
- Impact mitigation, including use of well-engineered systems.

These compensation methods do implicitly incorporate mitigation and insurance in several different forms and phases of the development

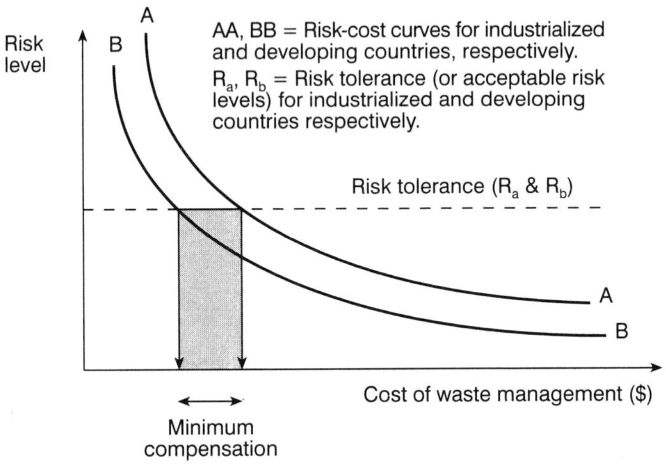

Fig. 6.3a Conceptual representation of adequate and equitable compensations for waste recipient countries: the case for equal risk tolerance levels.

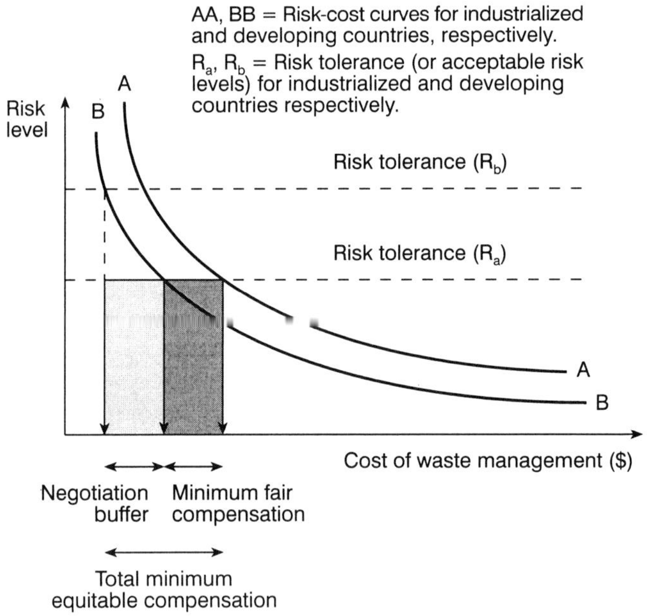

Fig. 6.3b Conceptual representation of adequate and equitable compensations for waste recipient countries: the case of different risk tolerance levels or 'double standards'.

programme. Indeed insurance should become an integral part of the hazardous waste management strategy. That is, further to the compensation, insurance coverage may be required, since it is believed that anticipation of the net liabilities remaining, and the desire to maintain continuing coverage generally provide substantial incentives for responsible persons and industry to engage in proactive risk management programmes and also to utilize safer disposal methods (Doherty *et al.*, 1990).

Ultimately, environmental precautionary measures are essential for continued economic development in the long term, and no amount of economic incentives proposed to compensate well-informed local residents may significantly influence their opinions on acceptance or rejection of hazardous wastes into their communities. When necessary, however, policy-makers should consider the equitable distribution of benefits and costs in making decisions to trade in hazardous wastes. The objective will be to extend the QALE as much as possible – given the extra funds available from potential waste trade programmes. Among other things, it is generally believed that improved nutrition, better sanitation, adequate health care and good education tantamounts to improved standards of living and/or quality of life. An assumption here is that society will, on average, use additional income at its disposal in a manner that reduces health and safety risks and therefore improves quality of life and increases LE.

Finally, it must be acknowledged that the current situation of wide variations in waste disposal costs in different countries promotes the transboundary movements of wastes in search of lowest-cost legal disposal. Consequently, 'low-cost disposal can undermine waste reduction and recycling efforts' (Long, 1990). Free trade in hazardous wastes may therefore not be a good policy for the international community – at least, not until the regional price disparities in disposal costs are bridged. The gaps in pricing for disposal costs can be closed by adopting adequate compensation packages for the waste recipient countries, even if disposal costs would have been extremely low in this case. This can be achieved by developing a framework to use in establishing adequate compensation levels of remittance to waste recipient countries in hazardous waste trade programmes. In any case, the apparent relaxation of conditions for court suits and the increase in the size of awards under appropriate conditions in some countries seem to confirm an apparently increased social emphasis on compensating those affected by toxic wastes (Doherty *et al.*, 1990).

A conceptual formulation for waste trade evaluations

7

Decision analysis concepts can be effectively used in the trade-offs analysis of a waste trade programme. As part of the underlying principles in its application, it is expected that a waste importing country will obtain some net financial gains and therefore increased GNP and GDP, which will then contribute to the overall well-being of the population in the country. On the other hand, potential environmental degradation and health risks may detract from any value-utility gained. An acceptable balance between the positives and the negatives is important; in particular, long-term consequences should not be overlooked by focusing only on potential short-term gains.

In general, curbing poverty, improving the quality of life, and safeguarding the environment for any country must present mutually supportive objectives (World Bank, 1989). A risk-cost-benefit technique may generally be employed to evaluate the extent to which a new policy can be used to arrive at a balance between economic development and environmental sustainability, and in particular, to improve the overall well-being of populations potentially affected by the development programme. Where associated risks exceed established international or relevant local standards, effective corrective action must be taken to reduce these risks to the acceptable levels.

7.1 A FRAMEWORK FOR TRADE-OFFS ANALYSIS

Any trade or industrialization policy, as well as the evolution of environmental management programmes is best analysed by taking account of and balancing all anticipated detrimental impacts against potential positive attributes by using a trade-offs or similar analysis. The basic philosophy is based on a utilitarian approach of balancing costs and benefits;

cost and benefit in this context are measured according to impacts on human well-being or welfare (or utility) – where well-being and welfare are measured by what society is willing to pay in the market-place for benefits, or what it is willing to accept by way of compensation for any costs (Pearce and Brisson, in Hester and Harrison, 1995).

To conduct a trade-offs analysis for a hazardous waste trade programme, the overall expected net benefits of the trade activities may be evaluated according to the following simplistic conceptual relationship:

$$ENB = \{MOC + \Sigma P_i B_i\} - \{\Sigma P_j C_j + TOC\},$$

where:

ENB = Expected net benefits from programme implementation
MOC = Monetary and other compensations to region importing wastes
P_i = Probability that benefit *i* (e.g. employment, improved domestic waste management, etc.) will be realized by region importing wastes as a result of new policy
B_i = Value of benefit *i* within region due to the given activity
P_j = Probability of detrimental impact *j* (e.g. reduced property values, health/environmental costs, etc.) in region due to the given activity
C_j = Costs of adverse impact *j* in performing the activity due to new waste management policy (including importing wastes)
TOC = Fixed transport, storage, disposal, long-term monitoring and administrative/regulations costs anticipated from the activity and/or for the new policy.

An important objective with respect to compensation and other benefits is to maximize ENB with the important constraints of minimizing P_j and C_j. P_j can be reduced by mitigative measures or protection, other kinds of abatement policies can be used to reduce C_j. Conditional compensations or payments such as insurance may also be used to reduce C_j, while in-kind compensations may increase B_i with respect to non-monetary benefits. A criterion should be involved, by which a decision to proceed with programme implementation makes some people better off and no one worse off. Beyond that, there should be equitable distribution of net benefits. Whereas it may not be possible to quantitatively estimate all the variables involved, it is important to recognize the importance of all variables and to perform at least some qualitative evaluation of all parameters involved.

The important variables to consider in a trade-offs evaluation for a waste trade programme or a hazardous waste management programme are indicated in Box 7.1. The trade-offs analyses result in the following:

$$\text{Total Net Benefits, TNB} = \Sigma\,[\text{Benefits}] - \Sigma\,[\text{Risks + Costs}]$$

$$\text{Net Benefits per Person} = \frac{[\text{Net benefits (in commensurate units)}]}{[\text{Population}]}.$$

If, on average, the net benefit is positive, then the programme may be acceptable and is therefore worthy of further consideration (provided risk tolerance criteria are met); if, on average, the net benefit is negative, then the programme should be rejected outright. Some of the relevant parameters are subjective and have high degrees of uncertainties associated with them. An applicable and appropriate evaluation framework is presented in Figure 7.1; this will facilitate the process of balancing the risks, costs and benefits for a given waste management or waste trade programme. Trade-off decisions made in the process will be directed at improving both short- and long-term benefits of the overall economic development programme.

7.2 TOWARDS AN INFORMED DECISION-MAKING

In attempts to arrive at informed decisions for development programmes, there is always the need to draw a balance between socio-economic development and environmental sustainability. For developing countries

Box 7.1
Important basic variables in a risk-cost-benefit trade-offs evaluation for waste management and waste trade programmes

Fundamental Parameters
- Population.
- Gross domestic product (GDP).
- Life-expectancy at birth (LE).
- Risk acceptance and tolerance criteria.

Risks and Costs
- Public health costs (e.g. medical care, loss in life-expectancy).
- Environmental costs (e.g. corrective actions, mitigation, pollution abatement).
- Safety/risk factors (e.g. facility integrity, monitoring and surveillance programmes).

Benefits
- Compensations (namely monetary, infrastructure, insurance, etc.).
- Gain in life-expectancy (GLE) or improved QALE.
- GDP growth rate.
- Improved management programmes for domestic wastes and environment.

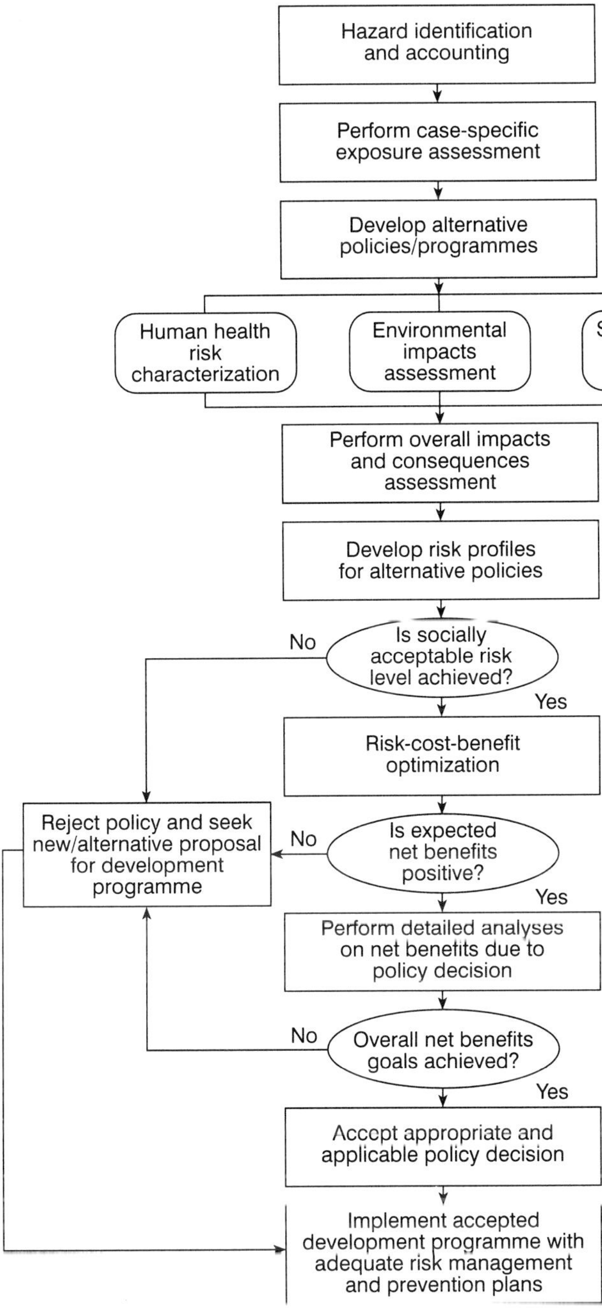

Fig. 7.1 A conceptual framework for risk-cost-benefit trade-offs analyses in a waste management/trade programme.

in particular, the key issue in their choice of a pattern or mode of growth (that will yield a combination of high economic gains with low environmental risks) is the assurance of an acceptable balance between economic growth and environmental sustainability.

Industrialization activities taking place in several countries will generally bring economic benefits; but most of these same activities carry potential health and environmental problems with them. In a similar manner, waste trade activities can bring both positive and negative effects to a region. Risk assessment and risk management methodologies – which have historically been used and found effective in evaluating similar problems, and also in making regulatory and policy decisions – can be employed to evaluate the impacts of such activities involving a hazardous waste trade plan. In this case, the degree to which qualities of life are placed at risk in undertaking a development activity can serve as a measure of its worth (Edgerton *et al.*, 1990) – or perhaps its worthlessness. This means that strategies for balancing the benefits with any trade-offs should be blended in the overall development programme. The balancing activity becomes feasible if one can consistently and effectively account for the gains in LE versus losses in LE, or its surrogate measures of net QALE or LP, or indeed the expected net benefits of a programme from the new policy.

To formulate a workable waste trade programme, where warranted, projected economic gains must be balanced against potential health and environmental risks and costs, even if only qualitatively. The focus would then be to seek an improved QALE or LP in an optimal way. Fundamental risk management concepts and applicable economic principles of cost-benefit analysis and multi-attribute utility theory can generally be used to test and evaluate this optimization hypothesis. A risk-cost-benefit analysis will help to compare the economic benefits with the associated risks and costs to health and the environment – with the overall goal of concurrently maximizing benefits and minimizing risks and costs. The evaluation can then serve as a basis for developing a framework for negotiations between the potential trading partners. In general, the level of risk should determine the type of response option to select in the decision problem.

It is noteworthy that economic development and environmental protection are not necessarily unalterably opposed choices. In fact, it has been noted that sensible policies and actions that protect the environment can at the same time contribute to economic progress (Schramm and Warford, 1989). Economic development and environmental protection should indeed complement each other; thus, improving one has the potential to enhance the other – as reflected by a World Bank observation that environmental problems generally are inseparable from the problems of development, and long-term economic growth depends on protecting the environment (World Bank, 1989).

7.3 LOGISTICAL INPUTS TO THE DECISION-MAKING PROCESS

For those countries wishing to make informed decisions about waste management and waste trade programmes there is plenty of useful data available for them to consult. Two of the most important sources of information are the IRPTC and UNEP's Computerized Environmental Law Information Base (CELIB).

IRPTC. In 1972 the United Nations Conference on the Human Environment, held in Stockholm, recommended the setting up of an international registry of data on chemicals likely to enter and damage the environment. Subsequently, in 1974, the Governing Council of the UNEP decided to establish both a chemicals register and a global network for the exchange of information contained in the register. An important aspect of the register's objectives was to assemble information on existing policies for control and regulation of hazardous chemicals at national, regional and global levels. In 1976 a central unit for the register, named the International Register of Potentially Toxic Chemicals, was created in Geneva, Switzerland, with the main function of collecting, storing and disseminating data on chemicals, and also to operate a global network for information exchange. IRPTC network partners (namely, the designation assigned to participants outside the central unit) consist of National Correspondents appointed by governments, national and international institutions, national academies of science, industrial research centres and specialized research institutions.

Chemicals examined by the IRPTC have been chosen from national and international priority lists. The selection criteria used include the quantity of production and use, the toxicity to humans and ecosystems, persistence in the environment and the rate of accumulation in living organisms. IRPTC stores information that would aid in the assessment of the risks and hazards posed by a chemical substance to human health and the environment. The major types of information collected include that relating to the behaviour of chemicals and information on chemical regulation. Information on the behaviour of chemicals, is obtained from various sources such as national and international institutions, industries, universities, private databanks, libraries, academic institutions, scientific journals and UN bodies such as the IPCS. Regulatory information on chemicals is largely contributed by IRPTC National Correspondents.

Following the successful implementation of the IRPTC databases, a number of countries went on to create National Registers of Potentially Toxic Chemicals (NRPTCs) that are completely compatible with the IRPTC system. The IRPTC, with its carefully designed database structure, provides a sound model for national and regional data systems. More importantly, it brings consistency to information exchange procedures within the international community. The IRPTC serves as an essential international tool for chemical hazards assessment, as well as a mechanism for information exchange on several chemicals. The establishment of any comprehensive monitoring and control network for the international trade in hazardous wastes will greatly benefit from the IRPTC system that is already in place. In fact, the IRPTC should be considered as an important logistical component in making policy decisions on hazardous waste trade programmes.

CELIB. CELIB which came online in January 1995, provides worldwide access to updated information on environmental law through the Internet – the most widely used international network communication service. Regarding international environmental conventions, CELIB contains a register of international treaties and other agreements, as well as texts of international conventions or agreements in the field of the environment. In addition, detailed information is provided in respect to several major global and regional conventions. Further information is provided in regard to national environmental legislation and institutions. This information is accessible by connecting through Gopher with 'UNEP.UNEP.NO' and selecting 'UN Environment Programme'.

CELIB should indeed be considered as an important logistical component in the making of policy decisions on hazardous waste trade programmes.

Assessing the viability of a waste trade programme

8

Like any other economic activity, a hazardous waste trade must be managed in an effective manner. Effective management requires that the risks associated with the various hazardous waste TSDFs are fully understood by all parties in any waste trade agreement, and also by all signatories to any relevant and related policies. Furthermore, the compensations from a waste trade programme must reflect the full costs of treatment, storage, disposal and long-term monitoring programmes to be borne by the waste recipient country. In addition, an acceptable waste trade should generate significant economic activities and economic growth for the waste importer. In fact, if it is decided that a waste trade must proceed, the waste recipient country should negotiate for net compensations that amount to at least what constitutes the net savings for the waste exporter.

8.1 A PROTOCOL FOR EVALUATING THE VIABILITY OF A WASTE TRADE PROGRAMME

An important question relating to the assessment of the viability of waste trade programmes is: What issues should be considered by a proposed waste recipient country before a final decision is made on any proposed waste trade activity? For example, for the developing and newly industrializing countries to realize the full benefits of any economic development activities (which may include waste trade programmes), good management policies and decision frameworks should be adopted in the policy-making processes. The appropriate framework or decision tool may incorporate conducting a preliminary risk appraisal to determine whether potential receptors would be at risk as a result of the economic development activity. In this case, appropriate exposure scenarios should be developed so that realistic decisions can be made regarding the level of risk posed or anticipated from a proposed development programme. Subsequently, an overall risk-cost-benefit assessment and/or a trade-offs analysis will help determine appropriate policy decisions for the proposed activity or economic development programme.

Box 8.1 contains the important factors that should generally be considered in evaluating hazardous waste trade programmes. Indeed steps to be agreed upon by all parties entering into this type of contract should be analysed in detail to affirm the utility of the whole programme. Such an analysis will allow the waste importing country to produce more efficient and consistent risk reduction policies, and to stimulate economic growth without jeopardizing public health and the environment.

A protocol for evaluating the viability of a toxic waste trade will generally involve several elements in the decision flow process. Some or all of the elements may have to be reviewed in a given situation, but each specific problem should be evaluated as unique in a case-specific environment. For example, if a given waste transaction belongs to a low-risk category (e.g. wastes of potential concern only for aesthetic reasons) then a different level of detail in the protocol may have to be adopted, compared to a transaction which potentially belongs to a high-risk category (e.g. wastes involving chemicals known to possess the ability to cause cancer, birth defects, etc.). A generic solution pathway is elaborated in the flowchart shown in Figure 8.1. The results from such an evaluation

Box 8.1

Important factors affecting hazardous waste import decisions

- Waste types to be imported and compatibility with local wastes.
- Background risks due to current waste management practices in waste recipient country.
- Anticipated transportation risks from different modes of transport associated with the proposed waste trade.
- Potential health and environmental risks to populations (i.e. short- and long-term receptor impacts), including indirect health effects.
- Potential socio-economic impacts on waste recipient country (i.e. socio-economic indicators and compensations resulting in improved lifestyle, increased quality of life, and increased life-expectancy).
- Availability of environmentally sound control standards and regulatory framework.
- Projected incremental (or decreased) risks due to waste imports policy.
- Risk tolerance of society/country (i.e. economic and cultural factors versus risk acceptability in waste recipient country).

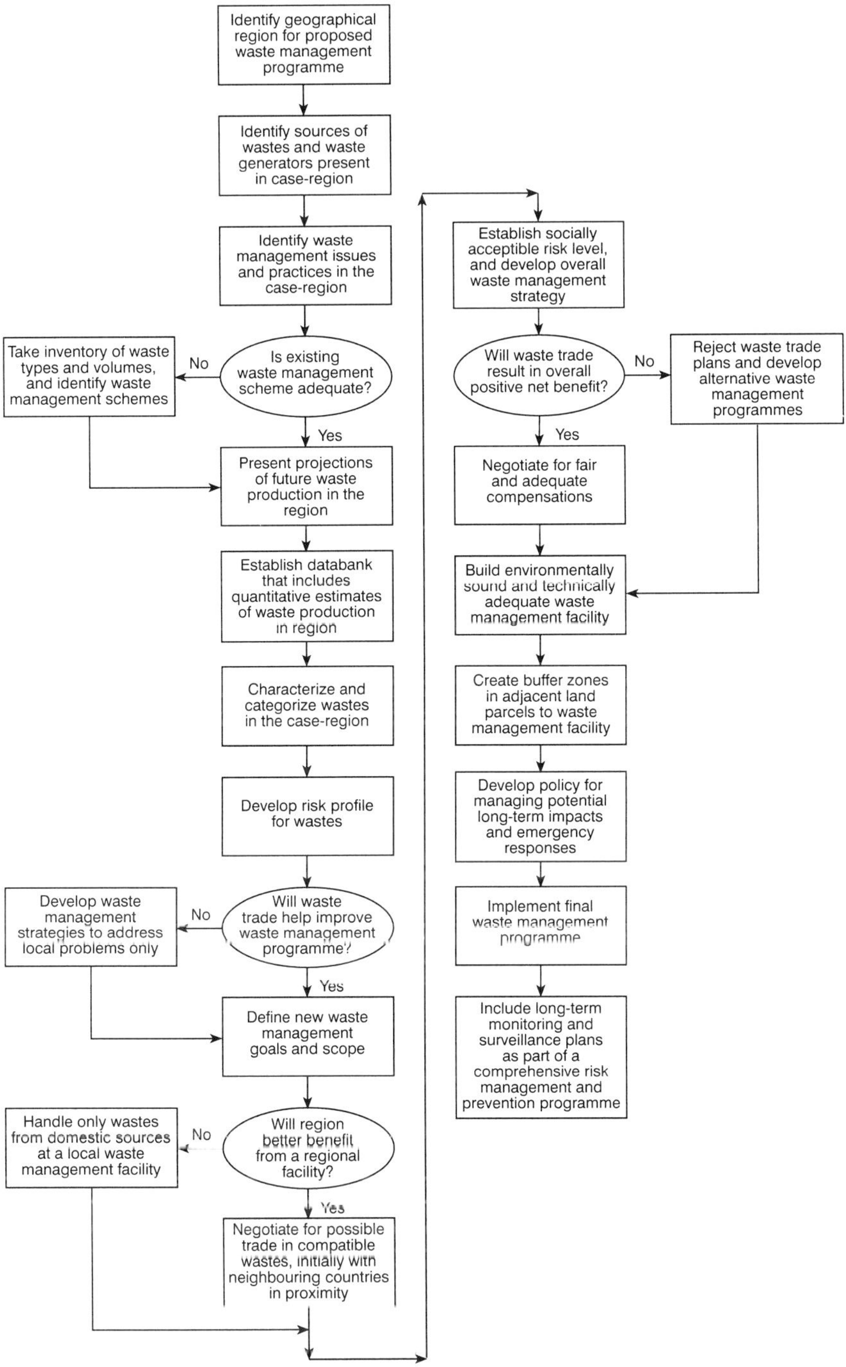

Fig. 8.1 Decision flowchart for evaluating the viability of waste trade programmes.

process will provide a context for balanced analysis in the decision-making, as well as produce enhanced development programmes that will be more acceptable to society.

8.2 JUSTIFICATION FOR A WASTE TRADE PROGRAMME

When could a waste trade be considered a reasonable policy? In answering this question, the argument can be made that, if the hazardous waste trade is managed efficiently, such trade programmes can be used to improve the overall environmental status, curb poverty, and improve the quality of life and general welfare within a region; all these goals can be managed as compatible and desirable objectives. Specific arguments for supporting a regulated and informed waste trade programme are elaborated below.

Economies of scale from regional waste management facilities. Hazardous wastes in developing and newly industrializing countries present potentially greater problems than wastes produced/managed in comparable regions of the industrialized world, because the latter tends to have regulations in place to help abate potential detrimental impacts from such wastes. For most hazardous wastes potentially present in many emerging economy countries, there may be 'insufficient' quantities to economically justify provision of complete waste management facilities (e.g. high temperature incinerators and catalytic units) to handle special wastes in each of the countries in a region. However, regional and sub-regional facilities and waste management systems serving some or all of the countries in a region could prove viable. Under such circumstances, waste trade and/or waste exchange programmes should be viewed more positively. A ban on transboundary waste shipments could indeed limit the scope of regional waste management centres – especially in their treatment, recycling and disposal capabilities.

Alleviation of local waste management problems. Participation in a waste trade may help alleviate existing local waste management problems, since domestic economies may not have the financial resources and capability to handle domestic wastes and local waste management practices. This could then help minimize the 'status quo' risks in society, and therefore help improve public health. A ban on the transfrontier shipments of wastes could indeed – directly or indirectly – act to inhibit economic growth, sustainable development and social well-being within a given country. As a corollary, any plans to participate in a waste trade should preferably be tied in to local waste management programmes.

Better global environmental protection/management. There may be good environmental and economic reasons for some transfrontier movements of wastes. A complete ban on transboundary shipments of hazardous wastes may not necessarily be the most effective way of protecting human health and the environment, because some waste generators may not have the capacity locally to dispose of all waste types in a technically sound and economical way. Banning all waste exports could potentially increase the incidence of illegal dumping. For instance, persons with criminal intents may offload wastes at sea, and the financial gains to criminals are great, whereas the chance of apprehension is rather limited; this has happened before and could become more rampant. Thus a complete ban on the transboundary movements of wastes may not be a best policy option. However, waste exports should **not** be permitted unless an environmentally sound disposal of the candidate wastes can be ensured in the waste recipient country.

Transfrontier shipments of wastes are likely to continue for several reasons that may be considered justifiable – including the fact that the closest suitable facility for a waste may be in a neighbouring country, or special wastes are generated in such small quantities that they will not justify the operation of a treatment facility in every country in a region. Indeed an industrialized country may even agree to import and dispose of or treat 'difficult' wastes for an emerging economy country under bilateral agreements. Also waste management strategies can be improved when countries cooperate in regional waste management programmes; such efforts may incorporate transboundary movements of wastes. Under all circumstances, the critical issue is that the transboundary movements of wastes should be conducted in an acceptable and safe manner to protect all parties.

Overall, the waste trade may be considered acceptable if significant environmental and economic benefits can be derived such that a reduction in the 'status quo' risks in a waste importing country is realized. In fact, if and when the hazardous waste trade should become a more 'legitimate' business, it can also serve as a platform for governments of developing nations to negotiate with, and woo industrial giants and multinational corporations to relocate some industries to the developing economies, which will provide jobs and economic well-being to the local populations. On the other hand, with the maintenance of public health and safety being crucial, financial constraints may not by themselves be justifiable enough reasons for any of the developing and newly industrializing countries to be participating in the toxic waste trade.

8.3 THE CASE AGAINST WASTE TRADE ACTIVITIES

Although the Basel Convention and related treaties put restraints and preconditions on waste exports, which may help to minimize the indiscriminate dumping of wastes abroad, waste exports are likely to continue, and could even increase. From the perspective of a waste recipient country – especially for the emerging economies – there is growing concern about waste trade activities. As a reaction to what is essentially an uncertain and 'unregulated' situation, the weight of political opinion is for an outright ban on any hazardous waste trade programme. Specific reasons frequently suggested to justify such a ban are discussed below.

Exploitative nature of waste trade transactions. The developing and newly industrializing countries that tend to be potential recipients of wastes shipped from industrialized economies are seen as very vulnerable. For instance, in the recent past, some developing countries appear to have been exploited and taken advantage of, because of their weak economic situations. In fact, the Basel Convention does little to ensure that wastes are not necessarily shipped to the most vulnerable cash-starved economy, where low disposal fees may be achieved at the expense of increased health and environmental risks. Also the fees paid to such waste recipient countries for disposing of the wastes may not fully reflect the long-term health and environmental risks. Hence there is growing concern that the potential long-term health and environmental risks exceed any short-term economic gains.

Uncertainties about the efficacy of waste trade programmes. Despite their general good intentions, international conventions are symptomatic of the cross-purposes, especially in view of the varying interests between developed and developing countries concerning the hazardous waste trade. There is no guarantee that under these conventions, waste recipient countries will be fully informed of the risks involved in the activity, and neither are disposal procedures in such countries likely to be closely scrutinized to help minimize potential risks to populations and the environment.

General lack of vital infrastructure and support in waste recipient countries. With the existence of international waste trade markets, profit-seeking waste exporters are likely to take advantage of the lower land and labour prices in less-industrialized countries, by building and operating waste disposal facilities to standards approved in the industrialized countries. However, the adequacy of international standards (e.g. for landfills and incinerators) to protect human health and the environment is questionable. Such facilities may be even less satisfactory in the emerging economy countries, which usually lack other vital/supportive infrastructure such as hospitals, emergency response programmes, reliable communication technology, and well-funded environmental regulatory programmes.

Offer of loopholes for misuse, abuse, and illegal dumping. The Geneva-based UN agency, IRPTC, warns that waste exporters are attempting to overcome the outcry over waste dumping in the developing countries by persuading the receiving nations that imported wastes could be recycled or converted to valuable commodities, such as energy. Thus, legislation that allows transboundary movements of wastes designated for recycling has the potential to offer loopholes that waste traders can adapt and misuse, to enable them to continue with illegal dumping activities.

In general, free trade in hazardous wastes may not be the best policy option for the international community working towards global environmental sustainability. Significant trade in wastes – be it legal or illegal – may indeed tend to act against incentives for waste minimization and recycling locally. This means that it would be essential to close the disposal cost disparities if the waste trade is still needed on any justifiable basis; the gaps in pricing can be bridged by adopting adequate compensation amounts for refunds to countries importing wastes. Lesser disparities in disposal costs will possibly act as a disincentive for the global waste trade – especially with respect to waste shipments from industrialized nations in search of cheaper disposal alternatives in the developing countries.

8.4 SHOULD THE TOXIC WASTE TRADE BE SUPPORTED?

Many developing and newly industrializing countries face a number of environmental problems ranging from uncontrolled air emissions to discharge of untreated wastes into bodies of water, to the often indiscriminate land disposal of toxic wastes. Most countries respond to the degradation of their environment by enacting comprehensive laws and regulations in an effort to combat the situation. An important issue, however, is whether these laws and regulations are enforced or widely ignored. In any case, a strategic and important way to address most of the environmental problems in the developing and newly industrializing countries may involve improving waste management strategies by cooperating in regional waste management programmes – which will generally involve the transboundary movements of wastes. In fact, regional waste trade programmes involving the emerging economies should help their regions attain a more sustainable environment. The relevant regional waste management and waste trade programmes should stipulate, among other things, the following agenda.

- Development of a policy to handle waste in the country of origin, as far as possible. That is, management of waste materials should preferably be carried out in the country of origin of the wastes, unless there is a sound rationale to addressing a problem mutually with 'neighbouring states' through a waste trade programme. From an ideal technical viewpoint, wastes should indeed be handled by the most appropriate facility, irrespective of the relative locations.

- A rational consideration of the benefits to the environment, and the scale of the operation that makes it feasible to set up joint waste management schemes with neighbouring countries for the treatment and disposal of certain waste streams. This would be applicable only under special conditions and/or scenarios.
- Implementation of regional waste management programmes in an atmosphere of shared responsibility. An atmosphere of confrontation which could potentially dominate such programmes can be replaced by one of interaction and common responsibility, without ignoring the differences and conflicts of interest between parties involved.

It is noteworthy that regional waste trade schemes do indeed exist between several industrialized countries. For instance, Canada and the US are major trading partners in hazardous wastes – with high volumes of wastes frequently shipped between the two countries under a bilateral treaty. However, for the industrialized countries, the waste trade can be conducted in a relatively fair and responsible atmosphere – contrary to what could happen in the less-industrialized countries.

Overall, waste imports should become an option for preliminary evaluation in a development programme only if environmentally sound operating standards can be ensured. This means that the hazardous waste management facilities in the waste recipient country should meet technical requirements that use health-based environmental standards. Several peripheral factors (including the country's regulatory programme; the sophistication of the waste management technologies and operational procedures at the waste-receiving facility; the waste-receiving facility's compliance track-record; and even broader socio-economic and political stability factors) may have to be considered as part of the evaluation and decision-making process.

With due consideration given to the above, the toxic waste trade may be considered acceptable if significant environmental and economic benefits can be derived such that a reduction in 'status quo' risks is realized. Box 8.2 contains specific general conditions that will normally contribute to a waste trade programme being considered acceptable. In fact, it may be argued that waste trade programmes can be conducted safely, given enough investment, training and infrastructure. Consider, for example, a scenario whereby 'Waste Management Company Ltd' (WMC) located in England cannot find a suitable location for a hazardous waste incineration, treatment and landfill complex due to NIMBYism. A location elsewhere in a developing country is found, and this facility is built to the same standard as it would have been in the UK. It is operated under strict and open regulatory controls, such that its continued high standards of performance are verifiable. WMC makes its money from importing wastes to this TSDF where the operating costs are generally lower than in the UK, and the developing country avails itself of a facility for its own

Box 8.2
Conditions necessary for a waste trade programme acceptability

- Economic gains must be balanced against potential health and environmental impacts.
- Negotiated compensation amounts should significantly increase the country's GDP.
- On the average, actions taken should increase the welfare of at least one individual without reducing that of any others in the society/country.
- Special training needs should be met for customs officials with respect to transboundary waste shipments.
- Programmes should be so implemented to enhance and stimulate economic growth without jeopardizing public health and/or the environment.

hazardous wastes produced by its emerging economy – further to the creation of jobs and commerce for the region. This will probably represent a most ideal scenario and a most environmentally acceptable manner for the conduct of a waste trade activity.

8.5 RECOMMENDED GUIDELINES FOR PLANNING A WASTE TRADE ACTIVITY

If and when it is decided that the waste trade must proceed for any reason, then the programme should be systematically evaluated before it is implemented. The question is: What issues should be considered by the proposed waste recipient country before any final decision is made on such a proposition? A complete evaluation of a waste trade programme should consider several assessments and issues. The following recommendations and suggestions elaborated below are made in connection with any decision on potential waste trades programmes.

Justify rationale for the waste trade. In part, the rationale for accepting wastes into a country may be provided initially by confirming the presence locally of similar or compatible types of wastes. Subsequently, an assessment of overall benefits to local residents and the nation as a whole from the importation of wastes for disposal in the region should be conducted. Social considerations – including the psychological impacts on local populations – should also be incorporated in the overall programme evaluation.

Ensure safe waste management practices. The adequacy of disposal facilities in the waste importing country should be ensured. Allowance should be made so that exporters ensure that there is a written agreement with recipient and corridor countries, and that recipient countries have the technical and management capacity to handle the wastes properly. The waste management approach should meet international standards.

Establish national regulatory control programmes. There is the need to establish national regulatory control programmes, with appropriate legislation, regulations, ordinances and licences that will help protect human health and the environment from any mismanagement of hazardous wastes – both locally generated and imported. It is important that various regions agree to exchange information on the transboundary movements of hazardous wastes and other hazardous materials; furthermore, they should jointly institute and enforce pertinent environmental laws that go beyond just hazardous waste dumping.

Design regional legislative framework. Specified nations in a region should develop appropriate/pragmatic regulatory frameworks, processes and technical infrastructure to help manage hazardous materials use or disposal in the various countries in the region. The design of a regional legislative framework to enforce the proper handling of hazardous wastes is deemed essential – including controls on generation, transportation and disposal. Implementation of regional waste management programmes should be conducted in an atmosphere of shared responsibility.

Establish adequate legal and liability protection. As far as possible, conduct waste trade negotiations at diplomatic levels, to allow effective legal actions against culprits and generators involved in illegal dumping of wastes. Also, provisions should be made for insurance and liability so that adequate financial resources are available for accidents, emergency responses, clean-ups, etc. Illegally exported and problem wastes must be returned to the country of origination. Anyone breaching the regulations may not only be required to return the wastes, but could also be made to face severe legal charges, fines and even jail terms.

Establish adequate compensation levels. There is the need to develop a framework for establishing adequate compensation levels for hazardous waste trade and exchange programmes. Appropriate compensation for waste disposal must recognize the full cost of risks, remedial actions and other direct costs borne by the waste recipient country. Given the uncertainties associated with predicting future risks, it is important that insurance becomes an integral part of any compensation package. The availability of adequate insurance protection as well as other compensation and mitigative measures must be ensured.

In the absence of a waste trade ban, importing countries should have all the benefits of safe waste disposal – including provisions for insurance and long-term liability protection. The waste recipient countries should also obtain some immediate reasonable compensation and/or economic growth from this activity. Where the waste trade is warranted, exporting countries should ensure that importing and corridor/transit countries are all informed, and written consent granted by all in the waste trade. Recipient and corridor countries should all be warned of potential dangers – and corridor countries in particular, if they deem it necessary, should be given enough time to file restraints and justifications for non-approval under international judicial interventions.

It is proposed here that, the Basel Convention works towards a programme to 'deprivatize' waste trades (at least partially) so that, when necessary, such activities are conducted only between governmental institutions via an international waste trade centre under the auspices of a UN-appointed expert working group. The overseeing UN panel of experts can also ensure responsible leadership – by monitoring how governments use funds earned from such a programme to improve the well-being of a nation's citizens. Open forums and the operation of full information systems and facts about waste trade activity will typically help in this direction. A waste recipient country failing to follow 'responsible' programmes of waste disposal, etc., could then be denied permission to import wastes again in the future.

Furthermore, the waste trade should not be viewed as a 'one-way' affair, whereby only industrialized nations may export wastes to developing and newly industrialized countries. Under bilateral agreements, the reverse flow of wastes could also be useful to the emerging economies that lack adequate facilities to handle potentially hazardous wastes being generated locally. Thus, to enable efficient management in the short term, arrangements could be made for high-risk hazardous wastes originating from developing and newly industrializing countries to be exported to TSDFs in industrialized nations; bilateral agreements will allow future acceptance, through possible imports, of wastes in the reverse direction, once appropriate infrastructures and regulatory frameworks are in place in the developing and newly industrializing countries.

Towards a global environmental policy-making 9

It is apparent that the world faces a global crisis that cannot be ignored. Recognizing the urgency of the global environmental problems, therefore, strict international cooperation and vigilance should be advocated amongst all the nations of the world – even amongst those with conflicting agendas. The protection of the global environment against pollution and deterioration has indeed become a very important international policy issue in recent years. For most developing and newly industrializing nations, environmental degradation has become a particularly serious problem – especially over the past decade or two. Part of the degradation in the emerging economies can be attributed to the rapid industrialization in some countries, and to hazardous wastes dumped from elsewhere onto some others that are usually in economically deprived regions. The problems of the emerging economies should not necessarily be ignored by the developed countries (some of which may have, directly or indirectly, contributed to some of these very problems) as the developing nations' problem; rather, it should be seen as a critical global environmental issue to be addressed now, and with much urgency and concerted international efforts – albeit such efforts are likely to be hampered by sovereignty concerns.

9.1 FORGING GLOBAL ENVIRONMENTAL AGREEMENT BETWEEN 'UNEQUAL' PARTNERS

It is apparent that most current global environmental problems have been caused by economic development activities in the industrialized nations of the world, although some developing countries may have enjoyed certain limited benefits from the industrial activities that resulted in much of today's environmental degradation. Now, in their quest to improve the quality of life for their populations, several developing countries are also fast contributing to global environmental problems. Even so, most developing countries have often felt that environmental protection is an unaffordable luxury for them; economic development and growth frequently have been the main focus of attention in all of their important policy decisions on development programmes for these countries.

Considering that the developed and less-developed countries are unequal parties in their contributions to global environmental degradation problems, and in view of the nature of prevailing policies in some of the former nations, the less-industrialized countries tend to be sceptical of the industrialized states' efforts towards global environmental protection. Nonetheless, most developing countries now recognize the need for environmental protection, having come to the realization that the likely cost of environmental protection (i.e. preventative measures taken against environmental degradation) will generally be less than the cost to repair a damaged environment (i.e. curative or corrective action measures taken to restore the impacted/damaged environment). The developing countries argue, however, that it is unfair for them to have to bear the same burden for environmental protection as do the industrialized countries, since much of the present environmental degradation problems may be attributed to activities by the industrialized world in the creation of wealth for their nations. Thus, the developing countries further argue that the industrialized states should take substantial action and bear most of the cost needed to ameliorate the damage that has already been done to the environment (Nanda, 1995).

By and large, the global community has come to realize that the question is not whether action should be taken by all states, but how to assign equitable and fair levels of responsibility to the different nations. The developing countries argue – and reasonably so too – that treaties and multilateral agreements should not place the same burdens on all nations, but rather should recognize notions of equity and provide for 'special' treatment of developing economies, allowing such countries to meet much lighter burdens (Nanda, 1995). One approach to lightening the responsibilities of the emerging economies may consist of a call on the major banks and multinational institutions to negotiate so-called 'debt-for-nature swaps', which allow developing nations to reduce their crippling debt burdens by setting aside and protecting precious natural resources (e.g. rainforests, ecologically sensitive preserves, etc.) that may help sustain a global ecological balance.

9.2 AN AGENDA FOR GLOBAL ENVIRONMENTAL PROTECTION

Many present day environmental problems can only be solved on the basis of international understanding, which is far from straightforward, given that adequate frameworks for cooperation are yet to be developed and/or implemented. In fact, the very nature of the environment, its complexity, its multiple interconnections and its in-built delay mechanisms render environmental problems difficult to understand and thus easy to ignore. This scenario has resulted in uncertainties and disagreements about the fields of greatest importance, and conflicts over boundaries of interest. Views vary between individuals and organizations, depending upon their particular concerns and interests, and also the time-scale of their outlook. As a consequence, there is still a scarcity of sound environmental information. Much of the relevant data are still missing and those which do exist are not always known or accessible. In general, insufficient use is made of existing data, but also those available are often inadequate, spatially and temporally patchy, incomplete and inconsistent. The underlying cause of many of these problems is the absence of an appropriate institutional framework to address environmental concerns directly and fully. Furthermore, the fact that environmental problems tend to vary significantly from one region to another makes it more challenging to develop consistent strategies that are applicable to the global community.

The UNCED process has come to affirm that national environmental legislation and related institutions are among the critical elements in capacity building for sustainable development. In this regard, there is an increasing focus on long-term environmental planning with the objective to integrate it into overall national development programmes. The need to incorporate environmental planning into national socio-economic programmes is therefore widely recognized at this time, with the environmental impact assessment (EIA) process having become (since the 1970s) the predominant tool for such integration.

After the 1992 UNCED (held in Rio de Janeiro, Brazil), the explicit purpose of which was to reconcile economic development with environmental protection, the immense complexity of the task of achieving sustainable world development became clear to the international community. Of particular interest, Agenda 21 (which is a comprehensive statement of concerns, findings and recommended actions on specific environmental areas) fundamentally commits the UN Member nations to a 'cradle-to-grave' approach to environmental management. In any case, to ensure that resources are properly husbanded and the world's wealth more equitably distributed, there needs to be a radical reappraisal of the wasteful lifestyle that modern societies have grown accustomed to, especially in the industrialized countries.

The developed countries have a particularly important role and responsibility to establish policies that take better account of sustainability (CEC, 1993). Among other things, the following may be considered important justification for designing an effectual international programme for global environmental protection.

- In the framework of the appropriate national regulatory control programmes, there is a need to establish a commercial and industrial waste register within certain disposal areas. The relevant data may be collected for the drafting of various waste management plans or programmes.
- There is an urgent need for the establishment of regional centres for training and technology transfer in regard to the management of hazardous and other wastes, and also the minimization of waste generation, in accordance with the specific needs of different regions and sub-regions.
- There is a real need to establish an international emergency fund for compensation for damage resulting from the transboundary movements of hazardous wastes and their disposal.
- The development and international harmonization of risk-assessment procedures and protocols should present particularly important opportunities and challenges to the global community, by ensuring the implementation of consistent environmental management programmes.

Ultimately, an established environmental policy that is publicly available and understood at all appropriate levels is fundamental to attaining the requisite environmental management standard. The policy has to indicate commitments to: reduce waste, pollution and resource consumption; minimize environmental risks and hazard effects; design products with regard to their environmental effects throughout the life cycle; control the effects of raw material processing; and minimize the effects of new developments. Furthermore, a system of environmental management records should be established in order to demonstrate compliance with the requirements of the environmental management system, and to record the extent to which the planned environmental objectives and targets are being met.

9.2.1 MULTILATERAL COOPERATION ON GLOBAL ENVIRONMENTAL PROTECTION

The problem of the environment should be an object of both national and international cooperation; only this combination of efforts can contribute to an effective solution to the global environmental problems the world faces. What Ananichev (1976) noted over two decades ago – that international cooperation in the field of environmental protection is an objective necessity for the peoples of the world not so much as a form of rational division of labour, but as an inevitable consequence of the progressive deterioration of the biosphere – is still very true today.

Solutions to environmental problems do indeed require innovative methods of approach, as well as the cooperation and participation of all sectors of society for their implementation. This should include: regulation (through legal and economic instruments), control and management, international cooperation and agreements, and monitoring and assessment. In the implementation of a multilateral cooperation agenda, environmental problems will usually be divided into global, regional, sub-regional and national issues. Such problems may therefore be addressed by the application of macro and micro procedures to the various aspects of the issues involved.

9.3 HARMONIZING GLOBAL ENVIRONMENTAL POLICIES

Differences exist, and will probably always exist in environmental management policies adopted by various countries and/or regions, consistent with each nation's overall development and sustainability goals. For example, from the beginning of the 1970s, a strong divergence of policies appeared between Eastern and Western Europe, because environmental concerns were not receiving significant attention in Eastern European countries (Kara, 1992). Since environmental issues were given less priority, it is not surprising that the importation of hazardous wastes from Western Europe, for instance, was subsequently embraced by the East. However, by the end of the 1980s growing global environmental euphoria advanced the desire to pursue a pan-European approach to environmental restoration and improvement.

Among other issues, the safe storage and disposal of hazardous wastes/materials has become one of the particularly critical environmental issues in both the developed and developing worlds. In fact, recent years have seen the development of another aspect of the environmental movement, characterized by concerns that encompass both the national and international issues which affect the widespread problems faced by the environment. In particular, the disposal of hazardous wastes has received increasing attention in the search for strategies to help tackle the extensive problems of waste management. To this end, a new world conservation strategy known as 'Caring for the Earth' was published in 1991, elaborating the principles and actions necessary to attain a sustainable global society.

Notwithstanding the fact that potential environmental problems vary in dimension and logistics according to the degree of development and the level of industrialization of a nation, there is one fundamental need for all countries – better programmes and policies are required globally for environmental problems. In fact, unless the industrialized nations help curb and control environmental degradation problems (due to both domestic waste management problems and waste imports) in developing countries, the campaign for the latter countries to be full participants in global environmental protection will be in vain. Under such circumstances, no real purpose will be served by the deforestation/afforestation programmes, amongst several other measures, that the developed countries support and have been encouraging the less-industrialized countries to undertake. Eventually the consequences of extensive environmental degradation affecting several areas globally could threaten the security of all countries in the world.

9.3.1 POLLUTION ABATEMENT STRATEGIES AND POLICIES

Overall, most of the developed countries have made significant progress in reducing emissions and waste stream discharges into the environment. This level of environmental protection has been reached by imposing strict emission and disposal standards on environmental pollutants from production processes. However, by the very nature of the technologies currently available and the relevant regulations, complete abatement of pollution problems is not possible in production processes. It is noteworthy that, in the past few decades, environmental technology has gone through a rapid evolutionary process. But there still remains the need for an innovative approach to product development and design that allows environmental problems to be tackled at source in the production process.

For now, the development of an end-of-pipe directed environmental technology has catalysed a lot of process and product innovation, and many new technologies aimed at waste prevention are under way. Other than the apparent financial burdens suffered in attempts to meet environmental regulations, these cleaner technologies promise significant trade-offs in terms of improved product quality, enhanced production capacity, better process controlability, and increased product reliability. Unfortunately some industries still do not appreciate the fact that, due to the product- and process-specific nature of the cleaner technology approach, substantial environmental and economic benefits could be realized in the long term. As exemplified by a comparison of the economic parameters for operating an existing industrial plant versus the installation of a new plant, it becomes apparent that some industries refuse to look beyond the short term, and consider the introduction of new and improved facilities to be far too costly. As a result, existing plants with obsolete technology and high potential for pollution and hazardous waste generation remain in operation (Neubacher, 1988). Nevertheless, with the serious commitment by the global community towards harmonizing environmental policies, there are great opportunities for even the sceptics to come to appreciate the innovative strategies being pursued. After all, the process of cooperative regulation tends to lead to greater harmonization of international environmental laws and practices, and may contribute to reducing the inhibiting effects which conflicting regulatory regimes are likely to have on the development of appropriate policies.

9.3.2 GLOBAL ENVIRONMENTAL MANAGEMENT STANDARDS

As a new development, the world's first environmental management standard was introduced by the British Standards Institute in 1992. It specifies requirements for the development, implementation and maintenance of environmental management systems. In June 1993, a meeting held in Toronto agreed to develop international standards for environmental management within the International Organization for Standardization (ISO). Of particular interest, the ISO 9000 series governs worldwide quality assurance programme certification.

In what may be viewed as a partial attempt to restore/maintain public confidence in industry, the EU issued its Council Regulation No.1836/93 on the voluntary participation of commercial enterprises in a common environmental management and audit scheme (EMAS). The Regulation entered into force on 13 July, 1993, with its ultimate application having been dependent on the issuing of national regulations in Member States. In these regulations, the prevention, reduction and elimination of environmental burdens – possibly at their source – result in a sound management of resources on the basis of the 'polluter pays principle', and the promotion of market-oriented environmental management is required. These are meant to stimulate and foster responsibility among commercial and industrial enterprises. The core elements of the Regulation are the introduction of an environmental management scheme at specific industrial sites and the execution of an environmental audit by independent auditors. In contrast to the environmental impact assessment which is carried out before a facility is installed, environmental audits take place during plant operations and trade activities. After a successful participation in EMAS, the company obtains a participation certificate (comparable to the certificates for quality management schemes in keeping with ISO 9000). It is noteworthy that the major difference between EMAS and the ISO version is that EMAS requires a public statement – i.e. transparency in environmental terms.

In the arena of transboundary shipments of wastes, for example, there is a need for organizing an international waste trade management audit system which specifies: what is to be audited, the frequency of audit, auditing responsibilities, personnel requirements, the nature of the audit protocol, and procedures for reporting audit findings. A corresponding waste trade management review can reduce risk to the environment in both the developed and the developing countries. For waste managers, further benefits that may accrue from such a system include: meeting and anticipating environmental requirements, reducing uncertainty, meeting customer requirements, improving corporate image with the public, raising asset value and improving profitability by minimizing waste, and reducing energy and resource consumption.

9.4 POLITICAL AND DIPLOMATIC ISSUES IN HAZARDOUS WASTE POLICY DECISION-MAKING

Waste management issues appear to be at centre-stage in many countries' environmental concerns (OECD, 1994). Traditionally, society's responses have been mainly directed towards the collection, treatment and disposal of wastes. Increasingly, however, efforts are now being aimed at waste minimization, which has become a fundamental goal of emerging environmental policies; it can be achieved through waste prevention, recycling and recovery, and more broadly through a better integration of environmental concerns in consumption and production patterns (OECD, 1994).

There are several types of regulatory decisions that can shape or affect the implementation of hazardous waste programmes; such decisions are often affected by political and even diplomatic factors. As Davis (1993) notes, in the US and probably several other places, 'political criteria have clearly assumed more prominent role in the hazardous waste policymaking process than economic or technical factors'. Indeed hazardous waste policy problems can become entangled with and complicated by political as well as economic and technical factors; but the key concern is the degree of commitment of the policy-makers responsible for such decision-making.

Typically, regulatory agencies usually are left in rather delicate positions because of the need to balance technical and political factors in reaching a hazardous waste policy decision. A case in point concerns the siting of hazardous waste management facilities (i.e. TSDFs); this presents one of the most controversial aspects of hazardous waste policy-making. This is because the NIMBY and related syndromes continue to pose a significant political hurdle for policy-makers seeking TSDFs for certain regions. Whereas these forms of opposition are difficult to overcome under any set of circumstances, there are both political factors and institutional forms that may contribute to a less adversarial forum for site-selection decisions and the development of a policy consensus (Davis, 1993).

Added to all the political and institutional problems is the fact that a hazardous waste policy decision can become a foreign policy and diplomatic issue when there are transfrontier movements of wastes involved. A case in point relates to to the feud between some West African nations and certain OECD countries, cited previously in Chapter 4 (section 4.2.1). In fact, whereas many factors may contribute to conflicts amongst nations, it is now apparent that environmental degradation/stress problems could very well trigger a North-South/West-East conflict between the developed and developing countries. Environmental stress is a result of a complex set of processes that operate at different levels, and that have to be carefully identified and sorted out before a full understanding of relationships can be gained. This must be considered in economic, socio-political and developmental terms because it causes multiple and complex consequences that engender social disruptions, political chaos and potential regional/inter-state conflicts. As the environmental stress grows in different parts of the world, it has the potential to cause global insecurity and developmental crises. In this regard, inappropriate hazardous waste trade policies could accelerate environmental degradation so that, in the long term, they will engender social disruption and political instability.

An epilogue

10

Given the potential for high profits, it is likely that the market for hazardous wastes involving developing and newly industrializing countries will continue to be attractive and available. In fact, some observers have even argued that waste may eventually be seen as an important trade commodity, at which time the notion of a ban on hazardous wastes could be denounced as a non-tariff barrier or an effort on the part of the industrialized world to limit access of the emerging economies to a potentially 'revered' resource. Notwithstanding this, there is the potential for serious long-term negative impacts from waste trade activities; consequently, there is a need for a very careful evaluation of any national policy and development programme involving waste imports.

In developing possible waste trade programmes, it must be borne in mind that curbing poverty, improving the quality of life, and safeguarding the environment must be mutually supportive objectives, since environmental problems are basically inseparable from the problems of development. Long-term economic growth depends on protecting the environment, so that programmes should be enhanced to stimulate economic growth without jeopardizing public health and/or the environment. For workable waste trade programmes, economic gains must be balanced against potential health, environmental and socio-economic impacts. Compensation amounts should significantly increase a nation's socio-economic status. On average, actions taken should increase the welfare of at least one individual without compromising the welfare of any other individual or group in the society.

Ultimately, the best waste management strategies usually will consist of: reducing waste production per unit of GDP; adopting waste minimization targets; developing an information base on the sources, composition and management of various waste streams; establishing monitoring networks to ensure early warning of potential environmental contamination problems; developing inventories of potential contaminated sites, as a basis for setting priorities and planning for clean-up actions; harmonization of waste management standards amongst nations; and implementation of the principles established in the Basel Convention, and indeed other relevant treaties or supplementary agreements.

10.1 CONCLUDING REMARKS

If engagement in a hazardous waste trade programme can result in better management programmes for domestic wastes and the environment, improved health care, and improved living standards and quality of life without imposing excessive new risks, then it might be argued that it is a worthwhile activity to pursue within the overall objectives of desirable economic development.

In general, a regulated and controlled trade in hazardous wastes may be beneficial to both exporting and receiving countries, if conducted within an informed decision framework. However, the developing countries may not be ready to participate in this trade yet – at least not on a global scale in the international waste market-place. But a trade programme involving only the regional members of emerging economies will be beneficial, although it should be conducted in the most efficient manner possible.

In fact, an outright ban on the waste trade may be an unworkable position for the international community to adopt; it may be an overreaction to a poorly understood and ill-managed problem. A position should be adopted for the use of an approach that considers the hazardous waste trade under informed decisions. By using appropriate strategies in an informed atmosphere, risks can be minimized and net economic benefits realized from waste trade activities; this can be achieved in a manner that is fair, socially responsible and beneficial to the global community. But for now, the following noteworthy observations and recommendations may be helpful to the 'undecided' global community that is trying to come to terms with the international trade in hazardous wastes.

- A ban on all waste shipments from industrialized to less-industrialized countries may be appropriate and necessary at this point in time – albeit this should be considered a dynamic policy, which is subject to change at any time by consensus deliberations amongst the nations of the world. In fact, it may be a worthwhile effort for the appropriate national authorities to extend any trade ban on nuclear and industrial wastes to include **all** waste types (even if meant for recycling), until such a time that the waste trade is **irrevocably** considered a beneficial development programme.
- Relocating foreign firms to developing and newly industrializing countries should be carefully scrutinized and monitored – to include a

check on the firm's track record with respect to environmental compliance of its affiliates, and also branches of such firms in industrialized countries, where applicable.

- With the potential need for regional waste management facilities and joint environmental programmes, it is inevitable that a form of waste trade (on a cooperative and limited scale) continues globally – and that includes the emerging economy countries. In fact, the notion of economies of scale in hazardous waste management systems calls for the management of waste streams on a regional basis, when necessary, in order to promote a globally sound environmental management agenda.
- Stringent regulations alone do not guarantee a healthy environment; rather, cooperation between industry/commerce and government is a necessary and more important prerequisite for protecting and/or improving the global environment.
- Inappropriate economic policies and weak regulatory systems could lead to potentially serious environmental degradation. Continued environmental degradation will threaten the attainment of sustainable economic development and an improved quality of life anticipated from industrialization and/or waste trades. Since a degraded environment cannot sustain continued economic growth, economic development should be coupled with sound environmental management. Therefore, there is a real need for the integration of environmental and economic decision-making elements, as a means of contributing to sustainable economic development.
- 'Institutionalization' of the hazardous waste trade, as the Basel Convention is purported to do, may not by itself be undesirable, provided that guidelines fully protect the interests of all nations, and provided that compensation to waste recipient countries is fair and adequate.
- If a decision is made to proceed with a waste trade based on the projected net benefits, then it is important to ensure that international laws are in place to protect the importer in vulnerable developing countries in particular. It is worth remembering that waste policy issues can become more of a foreign policy or diplomatic issue – not just a health, environmental or even moral issue.
- The main ingredients of an effective waste trade programme for waste recipient countries (and even the world community) are equitable compensation, minimum risk, adequacy in management of domestic wastes and responsible management. A hazardous waste trade that is carried out without appropriate and fair compensations is indeed a poor economic development strategy.
- Rather than allow economic hardships to determine tolerable risk limits in waste management programmes, it is more acceptable to establish

risk constraints based on risk-cost-benefit evaluations; decisions involving hazardous waste disposal and/or trade can therefore be based on a trade-offs analysis. Under such conditions, a safe, and hopefully acceptable, level of risk can be established.

- The benefits and risks associated with a hazardous waste trade must be kept in perspective. The problems associated with it must be weighed against the existing, possibly more serious problems associated with toxic wastes generated locally by a designated waste recipient nation. In general, if additional revenue can be generated in the waste importing country so as to improve living standards and indeed increase LE, then the waste trade may be viable and worthy of further consideration.
- There is the need for financial and technical assistance to the developing and newly industrializing nations, to enable such countries to become full participants of the global environmental protection and improvement programmes. In particular, international cooperation in the provision and use of specialized waste management facilities seems appropriate if the global environment is to be adequately protected. Also, special training needs are required for several personnel to monitor and control any transboundary shipments of hazardous materials and wastes.
- It must be recognized that legislation to control hazardous wastes and the transfrontier movements of such wastes is not sufficient by itself, but rather much additional work regarding enforcement is required to make the control systems work in practice. In fact, legislation for the control of hazardous wastes is rather difficult to enforce unless adequate and effective facilities are available for the proper recycling, treatment or disposal of the wastes produced or imported.

All said and done, it may be reasonable to assume that TSDFs need to, and will ideally, be provided for the major part of hazardous wastes generated within a country. However, in real situations, there may be small quantities of special hazardous wastes for which the operation of such facilities in every country is not feasible or even not economically sensible. For such reasons, and under such circumstances, there will be the long-term need for the transfrontier movements of hazardous wastes, so that the special wastes can be exported to appropriate TSDFs in the proximity of the source of generation. Because some transfrontier shipments of hazardous wastes may be a long-term necessity for the international community, a complete ban of the hazardous waste trade may not be the best global policy to adopt; however, effective control procedures and enforcement mechanisms will be very important ingredients of all regulations controlling the international trade in wastes. Countries generating small quantities of hazardous wastes should indeed view the waste trade as an important economic option, as long as the waste recipients do not compromise their environmental sustainability. The critical issue is that transboundary movements of wastes should be conducted in an acceptable and safe manner to protect all parties.

Finally, the relationships between environmental policies and major economic development programmes should be better spelt out. This will allow more credible decisions to be made regarding the international trade in hazardous wastes. After all, progress does not consist simply of looking for a direction in which one must go indefinitely; true progress can also consist of looking for a place where one can stop and assess the benefits derived or derivable from transient and transitional environmental policies.

References

Acar, Y.B., Alshawabkeh, A.N. and Gale, R.J. (1993) Fundamentals of extracting species from soils by electrokinetics.' *Waste Management*, **13** (2), 141–51.

Adede, A.O. (1993) *International Environmental Law Digest*, Elsevier, Amsterdam.

Ananichev, K. (1976) *Environment: International Aspects*, Progress Publishers, Moscow.

Anderson, W.C. (ed.) (1993–95) *Innovative Site Remediation Technology*, Vols 1–8, American Academy of Environmental Engineers, Annapolis, Md.

ARB (Air Resources Board) (1991) Soil decontamination. Compliance Assistance Program, Air Resources Board, Compliance Division, Los Angeles, Calif.

Asante-Duah, D.K. (1991) The Economic and Environmental Implications of the Hazardous Waste Trade on West Africa. Policy Report to the McNamara Fellowships Program, Washington, DC. The Economic Development Institute, World Bank, Washington, DC.

Asante-Duah, D.K. (1993) *Hazardous Waste Risk Assessment*, CRC Press/Lewis Publishers, Boca Raton, Fla.

Asante-Duah, D.K. and Sam, P.A. (1995) Assessment of waste management practices in sub-Saharan Africa. *International Journal of Environment and Pollution*, **5** (2/3), 224–42.

Avogadro, A. and Ragaini, R.C. (eds) (1994) *Technologies for Environmental Cleanup: Toxic and Hazardous Waste Management*, Kluwer Academic Publishers, Dordrecht.

Batstone, R., Smith, Jr., J.E. and Wilson D. (eds) (1989) *The Safe Disposal of Hazardous Wastes – The Special Needs and Problems of Developing Countries, Vol. I, II & III.* A Joint Study Sponsored by the World Bank, the World Health Organization (WHO) and the United Nations Environment Programme (UNEP). *World Bank Technical Paper 0253–7494, No. 93.* World Bank, Washington, DC.

Bhatt, H.G., Sykes, R.M. and Sweeney, T.L. (eds) (1986) *Management of Toxic and Hazardous Wastes*, Lewis Publishers, Chelsea, Mich.

Blumenthal, Daniel S. (ed.) (1985) *Introduction to Environmental Health*, Springer Publishing, New York.

Burke, M. (1996) Health of a continent. *Environmental Science and Technology*, **30**, (4), 162A-67A.

Cairney, T. (ed.) (1987) *Reclaiming Contaminated Land*, Blackie Academic & Professional, Glasgow.

Cairney, T. (ed.) (1993) *Contaminated Land (Problems and Solutions)*, Blackie Academic & Professional, Glasgow/Chapman & Hall, London/Lewis Publishers, Boca Raton, Fla.

CEC (1992) Analysis of priority waste streams: Chlorinated solvents. *Information Document No. CEC-DG-XI*, Brussels.

CEC (1993) Proposal for a Council Directive on integrated pollution prevention and control. *Information Document No. C-311/06*, Luxembourg.

Charbeneau, R.J., Bedient, P.B. and Loehr, R.C. (eds) (1992) *Groundwater Remediation*, Water Quality Management Library, Vol. 8, Technomic Publishing Lancaster, Pa.

CIR (Center for Investigative Reporting) and Moyers, W. (1990) *Global Dumping Ground – The International Traffic in Hazardous Waste*, Seven Locks Press, Washington, DC.

Cohrssen, J.J. and Covello, V.T (1989) *Risk Analysis: A Guide to Principles and Methods for Analyzing Health and Environmental Risks*, National Technical Information Service (NTIS), US Dept. of Commerce, Springfield, Va.

Colopy, J.H. (1995) Poisoning the developing world: The exportation of unregistered and severely restricted pesticides from the United States. *Journal of Environmental Law*, **13**; 167–223.

Covello, V.T., Menkes, J. and Mumpower, J. (eds) (1986) Risk evaluation and management, in *Contemporary Issues in Risk Analysis*, Vol. 1, Plenum Press, New York.

Crepeau, T.E and Beltz, P.R. (1985) *European Network of Waste Exchanges*, Lewis Publishers, Boca Raton, Fla.

Davidson, A. (1990) *In the Wake of the Exxon Valdez: The Devastating Impact of the Alaska Oil Spill*, Douglas & McIntyre, Vancouver.

Davis, C.E. (1993) *The Politics of Hazardous Waste*, Prentice-Hall, Englewood Cliffs, NJ.

Doherty, N., Kleindorfer, P. and Kunreuther, H. (1990) An insurance perspective on an integrated waste management strategy, in, *Integrating Insurance and Risk Management for Hazardous Wastes*, (Eds Howard Kunreuther and M.V. Rajeev Gowda), Kluwer Academic Publishers, Boston, Mass, pp. 271–302.

Edgerton, S.A. *et al.* (1990) Priority topics in the study of environmental risk in developing countries: Report on a workshop held at the East–West Center, August, 1988. *Risk Analysis*, **10** (2), 273–83.

Egboka, B.C.E., Nwankwor, G.I., Orajaka I.P. and Ejofor, A.O. (1989) Principles and problems of environmental pollution of groundwater resources with case examples from developing countries. *Environmental Health Perspectives*, **83** (Nov.), 39–68.

Fischhoff, B, Lichtenstein, S., Slovic. P. *et al.* (1981) *Acceptable Risk*, Cambridge University Press, New York.

FMHE (Federal Ministry of Housing and Environment) (1982) *The State of the Environment in Nigeria*, Monograph Series No. 2, Solid Waste Management, Environmental Planning and Protection Division. (Work commissioned in 1981.)

Forester, W.S. and Skinner, H. (eds) (1987) *International Perspectives on Hazardous Waste Management – A Report from the International Solid Wastes and Public Cleansing Association (ISWA) Working Group on Hazardous Wastes*, Academic Press, London.

Garbutt, J (1995a) *Environmental Law*, 2nd edn, J. Wiley, Chichester.

Garbutt, J. (1995b). *Waste Management Law*, 2nd edn, J. Wiley, Chichester.

Gibbs, L.M. (1982) *Love Canal: My Story*, State University of New York Press, Albany.

Greenpeace (1988) *Database of Known Hazardous Waste Exports from OECD to non-OECD Countries*, Greenpeace, Washington, DC.

Greenpeace (1989) *Waste Trade Update*, **2** (3) (15 July), Greenpeace, Washington, DC.

Greenpeace (1993) *Toxic Trade Update*, Number 6.1, First Quarter 1993, Greenpeace, Toxic Trade Campaign, Washington, DC.

Greenpeace (1994) *Database of Known Hazardous Waste Exports from OECD to non-OECD Countries*, Greenpeace, Washington, DC.

Grisham, J.W. (ed.) (1986) *Health Aspects of the Disposal of Waste Chemicals*, Pergamon Press, Oxford.

Haas C.N. and Vamos, R.J. (1995) *Hazardous and Industrial Waste Treatment*, Prentice-Hall, Englewood Cliffs, NJ.

Harbison, J.S. and McLarty, T.L. (1996) A move away from the moral arbitrariness of Maquila and NAFTA-related toxic harms. *Journal of Environmental Law*, **14** (1); 1–39.

Hester, R.E. and Harrison, R.M. (eds) (1995) *Waste Treatment and Disposal*, The Royal Society of Chemistry, Herts., UK.

Hilz, C. (1990) *Toxic Waste Exports to the Third World: An Analysis of Policy Options and Recommendations for an International Protocol to Protect the Global Environment*. Center for Technology, Policy and Industrial Development, Paper No. HSMP 16, MIT, Cambridge, Mass.

Holmes, J.R. (ed) (1984) *Managing Solid Wastes in Developing Countries*, John Wiley, New York.

Jewell, T. and Steele, J. (1996) UK regulatory reform and the pursuit of 'Sustainable Development': The Environment Act 1995. *Journal of Environmental Law*, **8** (2), 283–300.

Jolley, R.L. and Wang, R.G.M. (eds) (1993) *Effective and Safe Waste Management: Interfacing Sciences and Engineering with Monitoring and Risk Analysis*, Lewis Publishers, Boca Raton, Fla.

Kara, J. (1992) Geopolitics and the environment: The case of Central Europe. *Environmental Politics*, **1** (2), 18–36.

Kerton, R.R. (1990) *Double Standards: Consumer and Worker Protection in an Unequal World*, The North–South Institute, Ottawa.

Kocher, D.C. and Hoffman, F.O. (1991) Regulating environmental carcinogens: where do we draw the line? *Environmental Science & Technology*, **25**, 1986–89.

Kublicki, N.M. (1994) The greening of free trade: NAFTA, Mexican environmental law, and debt exchanges for Mexican environmental infrastructure development. *Columbia Journal of Environmental Law*, **19** (59), 91–92.

Laurence, D. (1988) *Transfrontier Waste Shipments: A Priority Environmental Policy Issue for Europe? Hazardous Waste*, Elsevier Science Publishers, Amsterdam.

Lave, Lester B. and Upton, Arthur C. (eds) (1987) *Toxic Chemicals, Health, and the Environment*, Johns Hopkins University Press, Baltimore, Md.

Leonard, H.J. (1985) Confronting industrial pollution in rapidly industrializing countries: myths, pitfalls, and opportunities. *Ecology Law Quarterly*, **12**: 779–94.

Levine, A.G. (1982) *Love Canal: Science, Politics, and People*, Lexington Books, Mass.

Lind, N.C., Nathwani, J. and Siddall, E. (1990) Measurement of safety in relation to social well-being. Discussion Paper, Institute for Risk Research Paper No. 23, Waterloo, Ontario.

Lind, N.C., Nathwani, J.S. and Siddall, E. (1991) *Managing Risks in the Public Interest*, Institute for Risk Research, University of Waterloo, Ontario.

Lippitt, J., Walsh, J., Scott, M. and DiPuccio, A. (1986) Cost of remedial actions at uncontrolled hazardous waste sites: Worker health and safety considerations. *Project Summary Report No. EPA/600/S2–86/037.* US EPA, Washington, DC.

Lister, C. (1996) *European Union Environmental Law: A Guide for Industry*, J. Wiley.

Lönngren, R. (1992) *International Approaches to Chemicals Control*, Kemi, Stockholm.

Long, W.L. (1990) Economic aspects of transport and disposal of hazardous wastes. *Marine Policy International Journal*, **14** (3) (May), 198–204.

Massmann, J., and Freeze, R.A. (1987) Groundwater contamination from waste management sites: The interaction between risk-based engineering design and regulatory policy 1. methodology 2. results. Water Resources Research, **23** (2) (Feb.), 351–80.

Merk, David (1990) *The Maquiladora Program: Hazardous Wastes from Mexican Border Industries.* Proceedings of HAZMACON 90, Hazardous Materials Management Conference and Exhibition, April 17–19, 1990, Anaheim, Calif., Association of Bay Area Governments (ABAG), pp. 183–8.

Nanda, V.P. (1995) *International Environment Law and Policy*, Transnational Publishers, Irvington-on-Hudson, New York.

Nathwani, J., Lind, N.C. and Siddall, E. (1990a) *Risk-Benefit Balancing in Risk Management: Measures of Benefits and Detriments.* Presented at the Annual Meeting of the Society for Risk Analysis, 29th Oct.-1 Nov., 1989, San Francisco, Calif. Institute for Risk Research Paper No.18, Waterloo, Ontario.

Nathwani, J., Lind, N.C. and Siddall, E. (1990b) *Safety, Social Well-Being and Its Measurement.* Presented at the 1st World Congress on Safety Science, Sept. 24–26, 1990, Cologne, Germany. Institute for Risk Research Paper No. 21, Waterloo, Ontario.

Neubacher, F.P. (1988) *Policy Recommendations for the Prevention of Hazardous Waste*, Elsevier Science Publishers, Amsterdam.

Newsweek (1988) The global poison trade. 7 Nov., 66–8.

Nogueira, D.P. (1988) *Hazardous Chemicals and the Developing Countries*, Elsevier, Amsterdam.

NRC (National Research Council) (1994) *Alternatives for Ground Water Cleanup*, Committee on Ground Water Cleanup Alternatives, National Academy Press, Washington, DC.

Nyer, E.K. (1992) *Groundwater Treatment Technology*, 2nd edn, Van Nostrand Reinhold, New York.

Nyer, E.K. (1993) *Practical Techniques for Groundwater and Soil Remediation*, Lewis Publishers, Boca Raton, Fla.

OBG (O'Brien & Gere Engineers, Inc.) (1988) *Hazardous Waste Site Remediation: The Engineer's Perspective*, Van Nostrand Reinhold, New York.

OECD (Organization for Economic Co-operation and Development) (1985a) *Resolution of the Council on International Cooperation Concerning Transfrontier Movements of Hazardous Wastes*, C(85)100, OECD, Paris.

OECD (Organization for Economic Co-operation and Development) (1985b) *Transfrontier Movements of Hazardous Wastes – Legal and Institutional Aspects*, OECD, Paris.

OECD (Organization for Economic Co-operation and Development) (1986) *OECD Council Decision–Recommendation on Exports of Hazardous Wastes from the OECD Area*, C(86) 64(Final), OECD, Paris.

OECD (Organization for Economic Co-operation and Development) (1988) *Decision of the Council on Transfrontier Movements of Hazardous Wastes*, 685th Session, C(88)90(Final), OECD, Paris.

OECD (Organization for Economic Co-operation and Development) (1991) *Environmental Data,* OECD, Paris.

OECD (Organization for Economic Co-operation and Development) (1993) *Monitoring and Control of Transfrontier Movements of Hazardous Wastes,* OECD Environment Monograph No. 34, OECD, Paris.

OECD (Organization for Economic Co-operation and Development) (1994) *Environmental Indicators: OECD Core Set,* OECD, Paris.

O'Hare, M., Bacow, L. and Sanderson, D. (1983) *Facility Siting and Public Opposition,* Van Nostrand Reinhold, New York.

Opsal, Knut (1989) *The Export of Toxic Waste to Africa – What Do We Know,* AFTEN Technical Note No. 4, July, The World Bank, Washington, DC.

Paustenbach, D.J. (ed.) (1988) *The Risk Assessment of Environmental Hazards: A Textbook of Case Studies,* John Wiley, New York.

Peck, Dennis L. (ed.) (1989) *Psychosocial Effects of Hazardous Toxic Waste Disposal on Communities,* Charles C. Thomas Publishers, Springfield, Ill.

Pratt, M. (ed.) (1993) *Remedial Processes for Contaminated Land,* Institution of Chemical Engineers, War., UK.

Rebovich, D.J. (1992) *Dangerous Ground: The World of Hazardous Waste Crime.* Transaction Publishers, London.

Reed, S.C., Crites, R.W., and Middlebrooks, E.J. (1995) *Natural Systems for Waste Management and Treatment,* 2nd edn, McGraw-Hill, New York.

Rees, W.E. (1992) Ecological footprints and appropriated carrying capacity. *Environment and Urbanization,* **4** (2), 9–12.

Rowe, W.D. (1983) *Evaluation Methods for Environmental Standards,* CRC Press, Boca Raton, Fla.

Rowland, Anthony J. and Cooper, Paul (1983) *Environment and Health,* Edward Arnold, Sevenoaks.

Saccomanno, F.F., Shortreed, J.H., Van Averde, M. and Higgs, J. (1989) *Comparison of Risk Measures for the Transport of Dangerous Commodities by Truck and Rail.* Presented at the 68th Annual Meeting of the Transportation Research Board, January 1989, Washington, DC.

Schramm, G. and Warford, J.J. (eds) (1989) *Environmental Management and Economic Development,* a World Bank Publication, Johns Hopkins University Press, Baltimore, Md.

Schwartz, S.I. and Pratt, W.B. (1990) *Hazardous Waste from Small Quantity Generators – Strategies and Solutions for Business and Government,* Island Press, Washington, DC.

Shimmin, Kathleen G. (1990) *Emergency Planning and Response Along the US/Mexican Border.* Proceedings of HAZMACON 90, Hazardous Materials Management Conference and Exhibition, April 17–19, 1990, Anaheim, Calif., Association of Bay Area Governments (ABAG), pp. 231–7.

Sims, R. *et al.* (1986) *Contaminated Surface Soils In-Place Treatment Techniques,* Noyes Publications, Park Ridge, NJ.

Sims, R.C. (1990) Soil remediation techniques at uncontrolled hazardous waste sites, a critical review. *Journal of the Air & Waste Management Association,* **40** (5) 704–32 (May).

Stanners, D. and Bourdeou, P. (eds) (1995) *Europe's Environment, The Dobris Assessment,* European Environment Agency, Copenhagen.

Suter II, G.W. (1993) *Ecological Risk Assessment,* Lewis Publishers, Boca Raton, Fla.

Tedder, D.W. and Pohland, F.G. (eds) (1990) *Emerging Technologies in Hazardous Waste Management*, ACS Symposium Series 422, American Chemical Society, Washington, DC.

Tedder, D.W. and Pohland, F.G. (eds) (1991) *Emerging Technologies in Hazardous Waste Management II*, ACS Symposium Series 468, American Chemical Society, Washington, DC.

Tedder, D.W. and Pohland, F.G. (eds) (1993) *Emerging Technologies in Hazardous Waste Management III*, ACS Symposium Series 518, American Chemical Society, Washington, DC.

Tedder, D.W. and Pohland, F.G. (eds) (1994) *Emerging Technologies in Hazardous Waste Management IV*, ACS Symposium Series 554, American Chemical Society, Washington, DC.

Tedder, D.W. and Pohland, F.G. (eds) (1995) *Emerging Technologies in Hazardous Waste Management V*, ACS Symposium Series 607, American Chemical Society, Washington, DC.

Third World Network (1989) *Toxic Terror – Dumping of Hazardous Wastes in the Third World*, Third World Network, Penang.

Tolba, M.K. (1990) The global agenda and the hazardous waste challenge. *Marine Policy International Journal*, **14** (3) (May); 205–9.

Travis, CC, Richter, S.A. and Crouch, E.A.C. (1987) Cancer risk management. *Environmental Science & Technology*, **21**, 415–20.

Trebilcock, M.J. and Howse, R. (1995) *The Regulation of International Trade: Political Economy and Legal Order*, Routledge, London.

UNDP (United Nations Development Programme) (1990) *Human Development Report 1990*, Oxford University Press, New York.

UNEP (United Nations Environment Programme) (1989) Basel Convention on the Control of Transboundary Movements of Hazardous Wastes and their Disposal. Final Act. *UNEP Report No. UNEP/IG.80/3*, 22 March 1989, Basel, Switzerland.

UNICRI (1993) *Environmental Crime, Sanctioning Strategies and Sustainable Development*, UNICRI Publication No. 50, United Nations Interregional Crime and Justice Research Institute/Australian Institute of Criminology, Rome/Canberra.

USEPA (US Environmental Protection Agency) (1974) Safe Drinking Water Act. Public Law 93–523.

USEPA (US Environmental Protection Agency) (1984) *Review of In-place Treatment Techniques for Contaminated Surface Soils*, vols 1 and 2, EPA-540/2–84–003a and b, US Environmental Protection Agency, Hazardous Waste Engineering Research Laboratory, Cincinnati, Ohio.

USEPA (US Environmental Protection Agency) (1985a) *Modeling Remedial Actions at Uncontrolled Hazardous Waste Sites*, EPA/540/2–85/001 (Apr. 1985), Office of Emergency and Remedial Response, Washington, DC.

USEPA (US Environmental Protection Agency) (1985b) National Primary Drinking Water Regulations; Volatile Synthetic Organic Chemicals; Final Rule and Proposed Rule. *Federal Register*, **50**, 46830–46901. National Primary Drinking Water Regulations; Synthetic Organic Chemicals; Inorganic Chemicals and Microorganisms; Proposed Rule. *Federal Register*, **50**, 56936–47025.

USEPA (US Environmental Protection Agency) (1987) *The New Superfund: What It Is, How It Works*, US Environmental Protection Agency, Washington, DC.

USEPA (US Environmental Protection Agency) (1988a) *CERCLA Compliance with Other Laws Manual (Interim Final),* EPA/540/G-89/006, Office of Solid Waste and Emergency Response, Washington, DC.

USEPA (US Environmental Protection Agency((1988b) *Guidance on Remedial Actions for Contaminated Ground Water at Superfund Sites,* EPA/540/G-88/003, Office of Emergency and Remedial Response, Washington, DC.

USEPA (US Environmental Protection Agency) (1989a) *CERCLA Compliance with Other Laws Manual: Part II – Clean Air Act and Other Environmental Statutes and State Requirements,* EPA/540/G-89/009, OSWER Directive 9234.1–02, Office of Emergency and Remedial Response, Washington, DC.

USEPA (US Environmental Protection Agency) (1989b) *Soil Vapor Extraction VOC Control Technology Assessment,* EPA-450/4–89–017 (Sept. 1989), Office of Air Quality Planning and Standards, Research Triangle Park, NC.

USEPA (US Environmental Protection Agency) (1990) *Air Stripper Design Manual.* Air/Superfund National Technical Guidance Study Series, EPA-450/4–90–003 (May 1990), Office of Air Quality Planning and Standards, Research Triangle Park, NC.

US/Mexico Hazardous Waste Work Group (1989) *The Maquiladora Industries Hazardous Waste Management Manual,* 1st Edn. Distributed by the Border Trade Alliance and the National Maquiladora Association.

Vallette, J. (1989) *The International Trade in Wastes – A Greenpeace Inventory,* 4th Edn, Greenpeace USA, Washington, DC.

Vallette, J. and Spalding, H. (eds) (1990) *The International Trade in Wastes: A Greenpeace Inventory. International Waste Trade Schemes and Related International Policies, 5th edn.* Compiled by the Greenpeace International Waste Trade Project, Greenpeace, Washington, DC.

Vandegrift, G.F., Reed, D.T., and Tasker, I.R. (eds) (1992) *Environmental Remediation: Removing Organic and Metal Ion Pollutants,* ACS Symposium Series 509, American Chemical Society, Washington, DC.

Vidic, R.D. and Pohland F.G. (eds) (1995) *Innovative Technologies for Site Remediation and Hazardous Waste Management.* American Society of Civil Engineers, New York.

Vir, A.K. (1989) Toxic trade with Africa. *Environmental Science & Technology,* **23** (1), 23–5.

Wang, L.K. and Wang, M.H.S. (eds) (1992) *Handbook of Industrial Waste Treatment,* Vol. 1, Marcel Dekker, New York.

Weiss, E.B. (ed.) (1992) *Environmental Change and International Law: New Challenges and Dimensions,* United Nations University Press, Tokyo.

Whipple, C. (1987) De minimis risk, in *Contemporary Issues in Risk Analysis,* Vol. 2. Plenum Press, New York.

Wise, D.L. and Trantolo, D.J. (eds) (1994) *Remediation of Hazardous Waste Contaminated Soils,* Marcel Dekker, New York.

Wold, C. (1996) Multilateral environmental agreements and the GATT: Conflict and resolution? *Environmental Law,* **26**, 841–921.

World Bank (1989a) *Striking a Balance – The Environmental Challenge of Development,* IBRD/World Bank, Washington, DC.

World Bank (1989b) *World Development Report 1989,* Oxford University Press, Oxford.

Yakowitz, H. (1990) Monitoring and control of transfrontier movements of hazardous wastes: An international overview, in *The Management of Hazardous Substances in the Environment,* (eds K.L. Zirm and J. Mayer, Elsevier Science Publishers Ltd., London, 139–62.

Further Reading

Abbou, R. (1988) *Hazardous Waste: Detection, Control, Treatment,* Elsevier, Amsterdam.

Abrams, D.J. (1990) Regulating the international hazardous waste trade: a proposed global solution. *Columbia Journal of Transnational Law,* **28**, 801–45.

Acar, Y.B. and Daniel, D.E. (eds) (1995) *Geoenvironment 2000: Characterization, Containment, Remediation, and Performance in Environmental Geotechnics,* Vols 1 and 2, ASCE Geotechnical Special Publication No. 46, American Society of Civil Engineers, New York.

Ahmad, Y.J. (1985) *The State of the Environment in Africa,* African NGO's Environment Network, ANEN/ELC, UNEP, Nairobi.

Ahmad, Y.J., El Serafy, S.E. and Lutz, E. (1989) *Environmental Accounting for Sustainable Development,* World Bank, Washington, DC.

Ahmad, Y.J. and Sammy, G.K. (1985) *Guidelines to Environmental Impact Assessment in Developing Countries,* Hodder & Stoughton, London.

Akinluye, T.O. and Odeyemi, O. (1984) Human waste disposal and the faecal pollution of the River Niger delta waters. *Water International (Switzerland),* **9** (Spring), 37–41.

Alberti, M. (1992) *Minimization of Hazardous Waste in Western Europe: Policy Implementation and Harmonization,* MIT, Cambridge, Mass.

Alloway, B. J. and Ayers, D. C. (1993) *Chemical Principles of Environmental Pollution,* Blackie Academic & Professional/Chapman & Hall, London.

AMCEN (African Ministerial Conference on the Environment) (1987) Report of the African Ministerial Conference on the Environment on the Work of its Second Session. *UNEP Report No. UNEP/AEC.2/3.20,* July 1987, Nairobi.

Andelman, J.B., and Underhill, D.W. (1988) *Health Effects From Hazardous Waste Sites,* repr., Lewis Publishers, Chelsea, Mich.

Anonymous (1988a) Africa: The industrial world's dumping ground. *African Business* (July) 11.

Anonymous (1988b) Exports of dangerous wastes to Guinea Bissau and Benin. *Europe Environment Review,* **2** (2) (July), 37–8.

Arendt, F., Annokkee, G.J., Bosman, R. and van den Brink, W.J. (eds) (1993) *Contaminated Soil '93,* Vols I & II, Kluwer, Dordrecht.

Asante-Duah, D.K. (1990) Quantitative risk assessment as a decision tool for hazardous waste management, in, *Proceedings of 44th Purdue Industrial Waste Conference (May, 1989),* Lewis Publishers, Chelsea, Michi., 111–23.

Asante-Duah, D.K. (1996) *Managing Contaminated Sites: Problem Diagnosis and Development of Site Restoration*, John Wiley, Chichester.

Asante-Duah, D.K., Saccomanno, F.F. and Shortreed, J.H. (1992) The hazardous waste trade (can it be controlled). *Environmental Science & Technology*, **26**, (9) (Sept.), 1684–93.

Asante-Duah, D.K., Saccomanno, F.F. and Shortreed, J.H. (1993) Tradeoffs evaluation of a hazardous waste trade program. *Waste Management & Research*, **11**, 63–79.

Ashby, E. (1980) *What Price the Furbish Lousewort?* Proceedings of the 4th Conference on Environmental Engineering Education, Toronto.

Ashford N.A. and Ayers, C. (1985) Policy issues for consideration in transferring technologies to developing countries. *Ecology Law Review*, **12** (4).

Barnard, R. and Olivetti, G. (1990) Rapid assessment of industrial waste production based on available employment statistics. *Waste Management & Research*, **8** (2), 139–44.

Bates, D.V. (1994) *Environmental Health Risks and Public Policy*, University of Washington Press, Seattle, Wash.

Bean, M.C. (1988) Speaking of risk. *ASCE Civil Engineer*, **58** (2), 59–61.

Benavides, L. (1990) Attempt to conduct an industrial wastes inventory in Lima, Peru. *Waste Management & Research*, **8** (2), 158–60.

Bernabe-Riefkohl, A. (1995) To dream the impossible dream: globalization and harmonization of environmental laws. *North Carolina Journal of International Law and Commercial Regulation*, **20**, 205–29.

Birnie, P.W. and Boyle, A.E. (1995) *Basic Documents on International Law and the Environment*, Clarendon Press, Oxford.

Biswas, A.K. and Khosla, A. (eds) (1990) *Environmental Modelling for Developing Countries.* Published for the International Society for Ecological Modelling with the Support of the United Nations Environment Programme, Tycooly Publishing, London.

Blackman, Jr., W.C. (1996) *Basic Hazardous Waste Management*, 2nd edn, Lewis Publishers/CRC Press, Boca Raton, Fla.

Bohnenblust, H. and Pretre, S. (1990) Appraisal of individual radiation risk in the context of probabilistic exposures. *Risk Analysis*, **10** (2), 247–53.

Boyle, A.E. (ed.) (1994) *Environmental Regulation and Economic Growth*, Clarendon Press, Oxford.

Brown, Lester (ed.) (1989) *State of the World 1989 – A World Watch Institute Report on Progress Toward a Sustainable Society*, Norton, New York.

BSI (1992) *Specifications for Environmental Management Systems*, British Standards Institution, London.

Bullard, R.D. (1990) *Dumping in Dixie – Race, Class and Environmental Quality*, Westview Press, Boulder, Colo.

Bullard, R.D. (1994) *Dumping in Dixie*, 2nd edn, Westview Press/HarperCollins, Boulder, Colo.

Cairns, Jr., J. and Crawford, T.V. (eds) (1991) *Integrated Environmental Management*, Lewis Publishers, Chelsea, Mich.

CAP (Consumers' Association of Penang) (1984) Toxic chemical and hazardous waste management in Malaysia. *CAP Report No. 4*, Penang.

Carson, W.H. (ed.) (1990) *The Global Ecology Handbook – What You Can Do About the Environmental Crisis*, The Global Tomorrow Coalition, Beacon Press, Boston, Mass.

Cassese, A. (1986) *International Law in a Divided World*, Clarendon Press, Oxford.

Cassidy, K. (1996) Approaches to the risk assessment and control of major industrial chemical and related hazards in the United Kingdom. *International Journal of Environment and Pollution*, **6** (4–6) 361–87.

Castleman, B.I. (1979) The export of hazardous factories to developing nations. *International Journal of Health Sciences*, **9** (4), 569–606.

CEC (1994) *Chemical Risk Control*, ECSC–EEC–EAEC, Luxembourg.

Charnovitz, S. (1993) Environmentalism confronts GATT rules. *Journal of World Trade*, **27**, 37–53.

Chatterji, M. (ed.) (1987) *Hazardous Materials Disposal: Siting and Management*, Avebury, Gower Publishing, Aldershot.

Chiu, H.S. and Tsang, K.L. (1990) Reduction of treatment cost by using communal treatment facilities. *Waste Management & Research*, **8** (2), 165–7.

Corn, M. (ed.) (1993) *Handbook of Hazardous Materials*, Academic Press, San Diego, Calif.

Cranor, C.F. (1993) *Regulating Toxic Substances: A Philosophy of Science and the Law*, Oxford University Press, New York.

Crouch, E.A.C. and Wilson, R. (1982) *Risk/Benefit Analysis*, Ballinger, Boston, Mass.

Csuros, M. (1994) *Environmental Sampling and Analysis for Technicians*, Lewis Publishers/CRC Press, Boca Raton, Fla.

Curi, K. (ed.) (1985) *Appropriate Waste Management for Developing Countries*, Plenum Press, New York.

Daniel, D.E. (ed.) (1993) *Geotechnical Practice for Waste Disposal*, Chapman & Hall, London.

Daniels, S.L. (1978) *Environmental Evaluation and Regulatory Assessment of Industrial Chemicals*, 51st Annual Conference Water Pollution Control Federation (WPCF), Anaheim, Calif.

Davis, A.P. (ed.) (1993) *Hazardous and Industrial Wastes*. Proceedings of the Twenty-Fifth Mid-Atlantic Industrial Waste Conference, Technomic Publishing, Lancaster.

Dixon, J.A., Scura, L.F., Carpenter, R.A. and Sherman, P.B. (1994) *Economic Analysis of Environmental Impacts*, Earthscan Publications, London.

Ellis, B. and Rees, J.F. (1995) Contaminated land remediation in the UK with reference to risk assessment: Two case studies. *Journal of the Institute of Water and Environmental Management*, **9** (1) (February), 27–36.

Falk, R.A. (1992) Toward a world order respectful of the global ecosystem. *Environmental Affairs Law Review*, **19B**, **C**, 711–13.

Fedra, K. (1989) *Modeling Hazardous Chemicals: Software Tools for Environmental Risk Analysis*, Environtech, Vienna.

Freeman, H.M. and Harris, E.F. (eds) (1995) *Hazardous Waste Remediation: Innovative Treatment Technologies*, Technomic Publishing, Lancaster, Pa.

French, H.F. (1990) A most deadly trade. *World Watch*, **3** (4) (July/August), 11–17.

Frey A.E. (1989) International transport of hazardous wastes. *Environmental Science & Technology*, **33** (5), 509.

Galli, C.D. (1987) Hazardous exports to the Third World: the need to abolish the double standard. *Columbia Journal of Environmental Law*, **12** (1), 71–90.

Ganster, P. and Walter, H. (eds) (1990) *Environmental Hazards and Bioresource Management in the United States–Mexico Borderlands*, UCLA Latin American Center Publications, University of California, Los Angeles.

Gilpin, A. (1995) *Environmental Impact Assessment (EIA): Cutting Edge for the Twenty-First Century,* Cambridge University Press, Cambridge.

Glasson, J., Therivel, R. and Chadwick, A. (1994) *Introduction to Environmental Impact Assessment,* University College London, UCL Press, London.

Goh, K.S. (1990) Developing hazardous waste programmes in Malaysia. *Waste Management & Research,* **8** (2), 151–6.

Gore, A. (1993) *Earth in the Balance: Ecology and the Human Spirit,* First Plume Printing, Penguin Books, New York.

Grant, J.A. (1994) Assessing and managing risk in the public sector: an urban hazardous waste landfill. *Journal of Urban Affairs,* **16** (4).

Greenpeace (1992) *Toxic Trade Update,* Number 5.2, Second Quarter 1992, Greenpeace, Toxic Trade Campaign, Washington, DC.

Gregory, R. and Kunreuther, H. (1990) Successful siting incentives. *Civil Engineering,* **60** (4), 73–5.

Gunn J.M. (ed.) (1995) *Restoration and Recovery of an Industrial Region: Progress in Restoring the Smelter-Damaged Landscape Near Sudbury, Canada,* Springer Series on Environmental Management, Springer-Verlag, New York.

Hadlock, C.R. (1995) *Multinational Corporations and the Transfer of Environmental Technology to Developing Countries,* International Environmental Affairs, A.D. Little, Cambridge, Mass.

Handl, G. and Lutz, R.E. (eds) (1989) *Transferring Hazardous Technologies and Substances: The International Legal Challenge,* Graham & Trotman, London/Dordrecht/Boston.

Handley, F.J. (1989) Hazardous waste exports: a leak in the system of international legal controls. *Environmental Law Reporter,* **19** (4) (April), 10171–82.

Handley, F.J. (1990) Export of waste from the United States to Canada: the how and why. *Environmental Law Reporter,* **20** (2) (February), 10061–6.

Hannequart, J.P. (1985) *Identification of Responsibilities in Hazardous Waste Management,* Environment Directorate, OECD, Paris.

Hansen, P.E. and Jorgensen, S.E. (eds) (1991) *Introduction to Environmental Management,* Developments in Environmental Modelling, 18, Elsevier, Amsterdam.

Harding, A.K. and Holdren, Jr., G.R. (1993) Environmental equity and the environmental professional. *Environmental Science & Technology,* **27** (10), 1990–3.

Harland, David (1985) *The Legal Aspects of the Export of Hazardous Products,* International Organization of Consumers Unions, Penang.

Hasan, S.E. (1996) *Geology and Hazardous Waste Management,* Prentice-Hall, Englewood Cliffs, NJ.

Hazardous Waste Inspectorate (1985) *Hazardous Waste Management: An Overview,* Department of the Environment, London.

Helfenstein A. (1988) U.S. controls on international disposal of hazardous waste. *The International Lawyer,* **22** (3) (Fall), 779–89.

Helle, M. (1985) *Status of Norwegian Hazardous Waste Management.* Meeting of Nordic Countries, 1985, Oslo.

Hickey, R.F. and Smith, G. (eds) (1996) *Biotechnology in Industrial Waste Treatment and Bioremediation,* Lewis Publishers/CRC Press, Boca Raton, Fla.

Hickok, H. (1996). *Introduction to Environmental Law,* Delmar Publishers, Albany, NY.

Hilz, C. (1992) *The International Toxic Waste Trade,* Van Nostrand Reinhold, New York.

HSE (Health and Safety Executive) (1989a) *Quantified Risk Assessment – Its Input to Decision Making,* HMSO, London.

HSE (Health and Safety Executive) (1989b) *Risk Criteria for Land-Use Planning in the Vicinity of Major Industrial Hazards,* HMSO, London.

Huang, C.P. (ed.) (1994) *Hazardous and Industrial Wastes.* Proceedings of the Twenty-Sixth Mid-Atlantic Industrial Waste Conference. Technomic Publishing, Lancaster, Pa.

Huntoon, B.D. (1989) Emerging controls on transfers of hazardous waste to developing countries. *Law and Policy in International Business,* **21**, 247–71.

Hurrel, A. and Kingsbury, B. (eds) (1992) *The International Politics of the Environment,* Clarendon Press, Oxford.

HWR (Hazardous Waste Report) (1989) Administration waste export bill requires EPA consent, bilateral agreements. *Hazardous Waste Report,* **11** (1), 2–3.

ICRCL (Interdepartmental Committee on the Redevelopment of Contaminated Land) (1987) *Guidance on the Assessment and Redevelopment of Contaminated Land,* 2nd Edn, *ICRCL 59/83.* Department of the Environment, Central Directorate on Environmental Protection, London.

IRPTC (1985a) *Industrial Hazardous Waste Management,* Industry and Environment Office and the International Register of Potentially Toxic Chemicals, United Nations Environment Programme, Geneva.

IRPTC (1985b) *Treatment and Disposal Methods for Waste Chemicals: IRPTC File,* International Register of Potentially Toxic Chemicals, United Nations Environment Programme, Geneva.

ISEP (1990) *Hazardous Waste Management, Contaminated Sites and Industrial Risk Assessment.* Proceedings of the Envirotech, 1990, Vienna.

ISEP (1992) *Information Technology in Waste Management.* Proceedings of the Envirotech, 1992, Vienna.

Ives, J.H. (ed.) (1985) *The Export of Hazard – Transnational Corporations and Environmental Control Issues,* Routledge & Kegan Paul, Boston, Mass.

Jain, R.K., Urban, L.V., Stacey, G.S. and Balbach, H.E. (1993) *Environmental Assessment,* McGraw-Hill, New York.

Johnson, S.P. and Corcelle, G. (1989) *The Environmental Policy of the European Communities,* Graham & Trotman, London/Dordrecht/Boston.

Jones, W.F. (1993) The evolution of the Bamako convention: an African perspective. *Colorado Journal of International Environmental Law & Policy,* **4**, 324–42.

Jorgensen, E.P. (ed.) (1989) *The Poisoned Well – New Strategies for Groundwater Protection,* Sierra Club Legal Defense Fund, Island Press, Washington, DC.

Kebe, M. (1990) Waste disposal in Africa. *Marine Policy International Journal,* **14** (3) (May), 251–3.

Keeney, R.D. and Raiffa, H. (1976) *Decisions with Multiple Objectives: Preferences and Value Tradeoffs,* John Wiley, New York.

Keeney, R.L. (1984) Ethics, decision analysis, and public risk. *Risk Analysis,* **4** 117–129.

Keeney, R.L. (1990) Mortality risks induced by economic expenditures. *Risk Analysis,* **10** (1), 147–59.

Keith, L.H. (ed.) (1992) *Compilation of E.P.A.'s Sampling and Analysis Methods,* Lewis Publishers/CRC Press, Boca Raton, Fla.

Kelly, W.E., Dahab, M.F., and Bogardi, I. (1993) *Environmental Engineering Education in a Global Community.* ENTREE '93 Proceedings, Delft.

Kettlewell, U. (1992) GATT – Will liberalized trade aid global environmental protection?' *Denver Journal of International Law and Policy,* **21**, 55.

Kimball, L. (1990) *Forging International Agreement: Strengthening Inter-Governmental Institutions for Environment and Development,* World Resources Institute, Washington, DC.

King, Jr., M.L. (1968) *Where Do We Go from Here: Chaos or Community?* Beacon Press, Boston, Mass.

Kiss, A. and Shelton, D. (1991) *International Environmental Law,* Transnational Publishers, New York.

Kleindorfer, P.R. and Kunreuther, H.C. (eds) (1987) *Insuring and Managing Hazardous Risks: From Seveso to Bhopal and Beyond,* Springer-Verlag, Berlin.

Kovar, J.D. (1993) A short guide to the Rio Declaration, *Colorado Journal of International Environmental Law and Policy,* **4,** 119–22.

Knowles, P.C. (ed.) (1992) *Fundamentals of Environmental Science and Technology,* Government Institutes, Rockville.

Kreith, F. (ed.) (1994) *Handbook of Solid Waste Management,* McGraw-Hill, New York.

Kubasek, N.K. and Silverman, G.S. (1994) *Environmental Law,* Prentice-Hall, Englewood Cliffs, NJ.

Kummer, K. (1994) *Transboundary Movements of Hazardous Wastes at the Interface of Environment and Trade,* UNEP Environment and Trade Monograph No. 7, Nairobi.

Kummer, K. (1995) *International Management of Hazardous Wastes,* Clarendon Press, Oxford.

Kungskulniti, N. (1990) Public health aspects of a solid waste scavenger community in Thailand. *Waste Management & Research,* **8** (2), 167–71.

Kunreuther, H. and Rajeev Gowda, M.V. (eds) (1990) *Integrating Insurance and Risk Management for Hazardous Wastes,* Kluwer Academic Publishers, Boston, Mass.

Kwiatkowska, B. and Soons, A.H.A. (eds) (1993) *Transboundary Movements and Disposal of Hazardous Wastes in International Law,* Graham and Trotman/Martinus Nijhoff (Kluwer Academic Publishers), Dordrecht.

Lassare, S., Fedra, K. and Weigkricht, E. (1990) *Computer-Assisted Routing of Dangerous Goods for Haute-Normandie.* ASCE Specialty Conference, 1990, St. Louis, Mo.

Lave, L.B. and Upton, A.C. (eds) (1987) *Toxic Chemicals, Health, and the Environment,* Johns Hopkins University Press, Baltimore, Md.

LCA Sourcebook (1993) *A European Business Guide to Life-Cycle Assessment*, Sustain Ability Ltd. London.

Lee, J.A. (1985) *The Environment, Public Health, and Human Ecology – Considerations for Economic Development.* A World Bank Publication, Johns Hopkins University Press, Baltimore, Md.

Lifson, M.W. (1972) *Decision and Risk Analysis for Practicing Engineers,* Barnes and Noble, Cahners Books, Boston, Mass.

Lindsey, A.W. (1988) *Managing Hazardous Waste,* State-of-the-Art, Washington DC, US, EPA.

Linnerooth, J. and Davis E. (1987) *Government Responsibility for Risk Management and Hazardous Waste,* Springer-Verlag, Berlin.

Linthurst, R.A., Bourdeau, P. and Tardiff, R.G. (eds) (1995) *Methods to Assess the Effects of Chemicals on Ecosystems.* SCOPE 53/IPCS Joint Activity 23/SGOMSEC 10, John Wiley, Chichester.

Lippmann, M. (ed.) (1992) *Environmental Toxicants: Human Exposures and their Health Effects,* Van Nostrand Reinhold, New York.

Lohwongwatana, B., Soponkanaporn, T. and Sophonsridsuk, A. (1990) Industrial hazardous waste treatment facilities in Thailand. *Waste Management & Research,* **8** (2), 129–34.

Louka, E. (1994) *Overcoming National Barriers to International Waste Trade: A New Perspective on the Transboudary Movements of Hazardous and Radioactive Wastes,* Graham and Trotman/Martinus Nijhoff (Kluwer Academic Publishers), Dordrecht.

Lowrance, W.W. (1976) *Of Acceptable Risk:Science and the Determination of Safety,* William Kaufman, Los Altos, Calif.

Ludwiszewski, R. (1993) Green language in the NAFTA: reconciling free trade and environmental protection. *International Law,* **27**, 691.

MacKenzie, D. (1989) If you can't treat it, ship it. *New Scientist (UK),* **122**, 24–5.

MacNeill, J.W. (1985) *Policy Issues Concerning Transfrontier Movements of Hazardous Waste.* Transfrontier Movements of Hazardous Wastes, OECD, Paris.

Magraw, D.B. (ed.) (1991) *International Law and Pollution,* University of Pennsylvania Press, Philadelphia, Pa.

Makarim, N. (1990) Hazardous waste management in Indonesia. *Waste Management & Research,* **8** (2), 157–8.

Martin, E.J. and Johnson, Jr., J.H. (eds) (1987) *Hazardous Waste Management Engineering,* Van Nostrand Reinhold, New York.

Martin, W.F., Lippitt, J.M. and Prothero, T.G. (1992) *Hazardous Waste Handbook for Health and Safety,* 2nd edn, Butterworth-Heinemann, London.

Maurits la Riviere, J.W. (1989) Threats to the world's water. *Scientific American.* (Sept.), Managing Planet Earth, Special Issue.

McDonald, J. (1993) Greening the GATT: harmonizing free trade and environmental protection in the new world order.' *Environmental Law,* **23**, 397.

McGregor, G.I. (1994) *Environmental Law and Enforcement,* Lewis Publishers/CRC Press, Boca Raton, Fla.

Meadows, D.H. Meadows, D.L. and Randers, J. (1992) *Beyond the Limits: Global Collapse or a Sustainable Future?,* Earthscan Publications, London.

Meriluoto, S. (1994) Solutions rest in integrated responsibility. *EEE Bulletin* (Winter), Helsinki.

Moavenzadeh, F. (1984) *Global Construction and the Environment: Strategies and Opportunities,* John Wiley, New York.

Moeller, D.W. (1992) *Environmental Health,* Harvard University Press, Cambridge, Mass.

Monahan, D.J. (1990) Estimation of hazardous wastes from employment statistics – Victoria, Australia. *Waste Management & Research,* **8** (2),145–9.

Moore, A.O. (1987) *Making Polluters Pay – A Citizen's Guide to Legal Action and Organizing,* Environmental Action Foundation, Washington, DC.

Morris, P. and Therivel, R. (eds) (1995) *Methods of Environmental Impact Assessment,* University College London Press, London.

Moyers, B. (ed.) (1990) *Global Dumping Ground.* Documentary on PBS series *FRONTLINE.* Center for Investigative Reporting (CIR), San Francisco.

NAE (National Academy of Engineering) (1993) *Keeping Pace with Science and Engineering: Case Studies in Environmental Regulation,* National Academy Press, Washington, DC.

Nagy, I.V. (1977) Optimization of waste disposal system. *Periodica Polytechnca,* (4), Budapest.

Nagy, I.V. (1986) *Public Participation in Environmental Protection in Hungary.* Seventh Meeting of the World Commission on Environment and Development, 1986, Moscow.
Nagy, I.V. (1987) *The Problems of Hazardous Waste Management in Hungary.* General Lecture, World Conference on Hazardous Waste, 1987, Budapest, Hungary.
Nash, J.M. (1990) Evolution of hazardous (chemical) waste disposal facilities in Hong Kong. *Waste Management & Research,* **8** (2), 163–4.
Neff, Alan (1990) Not in their backyards, either: a proposal for a foreign environmental practice act. *Ecology Law Quarterly,* **17** (3) (Fall), 401.
NRC (National Research Council) (1991) *Environmental Epidemiology (Public Health and Hazardous Wastes),* National Academy Press, Washington, DC.
NRC (National Research Council) (1995a) *Improving the Environment: An Evaluation of DOE's Environmental Management Program,* National Academy Press, Washington, DC.
NRC (National Research Council) (1995b) *Science and Judgment in Risk Assessment,* National Academy Press, Washington, DC.
OAU (Organization of African Unity) (1989a) *Control of Transboundary Movements of Hazardous Wastes and Their Disposal in Africa,* CM/Plen/Draft/Res.23(L)Rev.1. July 1989, Addis Ababa.
OAU (Organization of African Unity) (1989b) *Draft Resolution on Hazardous Wastes Protection of the Environment and Sustainable Development,* AHG/Draft.Res.5. July 1989, Addis Ababa.
OECD (Organization for Economic Co-operation and Development) (1985) *Transfrontier Movements of Hazardous Wastes – Legal and Institutional Aspects,* OECD, Paris.
OECD (Organization for Economic Co-operation and Development) (1988) *Decision of the Council on Transfrontier Movements of Hazardous Wastes,* 685th Session, C(88)90(Final), OECD, Paris.
O'Sullivan, D. (1988) Third world has most chemical contamination. *Chemical & Engineering News,* (3 October), 8.
Otway, H and Peltu M. (eds) (1985) *Regulating Industrial Risks: Science, Hazards and Public Protection,* Butterworths, London.
Paasikivi, R. (1994) Towards a European standard. *EEE Bulletin* (Winter), Helsinki.
Patnaik, P. (1992) *A Comprehensive Guide to the Hazardous Properties of Chemical Substances,* Van Nostrand Reinhold, New York.
Pedersen, J. (1989) *Public Perception of Risk Associated with the Siting of Hazardous Waste Treatment Facilities,* European Foundation for the Improvement of Living and Working Conditions, Dublin.
Peirce, J. and Vesilind, A. (1981) *Hazardous Waste Management,* Ann Arbor Science Publishers, Mich.
Philp, R.B. (1995) *Environmental Hazards & Human Health,* Lewis Publishers/CRC Press, Boca Raton, Fla.
Piddington, K.W. (1989) Sovereignty and the environment. *Environment,* **31** (7) (Sept.), 18–20, 35–9.
Porteous, A. (1985) *Hazardous Waste Management Handbook,* Butterworth, London.
Postel, S. (1988) *Controlling Toxic Chemicals.* State of the World 1988, Worldwatch Institute, New York.
Poterfield, A. and Weir, D. (1987) The export of U.S. toxic wastes. *The Nation* **245** (3 Oct.), 325.

Prager, J.C. (1995) *Environmental Contaminant Reference Databook,* Van Nostrand Reinhold, New York.
Prendergast, J. (1993) Engineering sustainable development. *Civil Engineering,* **63** (10) (Oct.), 39–42.
Prins, G. (ed.) (1993) *Threats Without Enemies* Earthscan Publications, London.
Prins, G. and Stamp, R. (1991) *Top Guns & Toxic Whales: The Environment & Global Security,* Earthscan Publications, London.
Raufer, R.K. (1996) Market-based pollution control regulation: implementing economic theory in the real world. *Environmental Policy and Law,* **26** (4), 177–184.
Ravishankar, R. (1990) Opportunities and constraints for cement kiln incineration in Malaysia. *Waste Management & Research,* **8** (2), 117–21.
Repetto, R. (1994) *Trade and Sustainable Development,* UNEP Environment and Trade Monograph No. 1, UNEP, Nairobi.
Rhyner, C.R., Schwartz, L.J., Wenger, R.B. and Kohrell, M.G. (1995) *Waste Management and Resource Recovery,* Lewis Publishers/CRC Press, Boca Raton, Fla.
Richardson, M.L. (ed.) (1986) *Toxic Hazard Assessment of Chemicals,* Royal Society of Chemistry, London.
Richardson, M.L. (ed.) (1992) *Risk Management of Chemicals,* Royal Society of Chemicals, Cambridge.
Rose, E.C. (1989) Transboundary harm: hazardous waste management problems and Mexico's maquiladoras. *The International Lawyer,* **23** (1) (Spring), 223–44.
Ruckelshaus, W.D. (1985) Risk, science, and democracy. *Issues in Science & Technology* (Spring), 19–38.
Russell, M. and Gruber, M. (1987) Risk assessment in environmental policy-making. *Science,* **236**, 286–90.
Salcedo, R.N. Cross, F.L. and Chrismon, R.L. (1989) *Environmental Impacts of Hazardous Waste Treatment, Storage and Disposal Facilities,* Technomic Publishing, Lancaster, Pa.
Salzman, J. (1991) *Environmental Labelling in OECD Countries,* Paris, OECD.
Sand, P.H. (ed.) (1992) *The Effectiveness of International Environmental Agreements,* Grotius, Cambridge.
Santos, M.A. (1995) *Limits and Scope of Environmental Law,* Charles C. Thomas Publisher, Springfield, Ill.
Schissel, H. (1988) The deadly trade: toxic waste dumping in Africa. *Africa Report* (10 Sept.).
Schleicher, K. (ed.) (1992) *Pollution Knows No Frontiers: A Reader,* Paragon House Publishers, New York.
Schmidt, A. (1992) Transboundary movements of waste under EC law: the emerging regulatory framework. *Journal of Environmental Law,* **4** (1), 57–80.
Schoenbaum, T.J. (1992) Free international trade and protection of the environment: irreconcilable conflict? *American Journal of International Law,* **86**, 700–27.
Schulin, R., Desaules, A., Webster, R. and Von Steiger, B. (eds) (1993) *Soil Monitoring: Early Detection and Surveying of Soil Contamination and Degradation,* Birkhäuser Verlag, Basel.
Sebek, V. (ed.) (1990) Maritime transport, control and disposal of hazardous waste. Special Issue of the *Marine Policy International Journal,* **14** (3) (May).
Setterberg, F. and Shavelson, L. (1993) *Toxic Nation,* John Wiley, New York.
Shere, M.E. (1995) The myth of meaningful environmental risk assessment. *Harvard Environmental Law Review,* **19** (2), 409–92.

Sincero, A.P. and Sincero, G.A. (1996) *Environmental Engineering: A Design Approach,* Prentice-Hall, Upper Saddle River, NJ.
Singhroy V.H., Nebert, D.D. and Johnson, A.I. (eds) (1996) *Remote Sensing and GIS for Site Characterization: Applications and Standards,* ASTM Publication No. STP 1279, ASTM, Philadelphia, Pa.
Sittig, M. (1994) *World-Wide Limits for Toxic and Hazardous Chemicals in Air, Water and Soil,* Noyes Publications, Park Ridge, NJ.
Smet, M. (1990) Management and control of transfrontier waste movements in Belgium. *Marine Policy International Journal,* **14** (3) (May), 228–35.
Smith, J.T. (1993) The challenges of environmentally sound and efficient regulation of waste – the need for enhanced international understanding. *Journal of Environmental Law,* **5** (1), 91–107.
Smith, L., Means, J. and Barth, E. (1995) *Recycling and Reuse of Industrial Wastes,* Battelle Press, Columbus, Ohio.
Smith, L.A. and Hinchee, R.E. (1993) *In-Situ Thermal Technologies for Site Remediation,* Lewis Publishers/CRC Press, Boca Raton, Fla.
Smith, Jr., T.T. (1996) Regulatory reform in the USA and Europe. *Journal of Environmental Law,* **8** (2), 257–82.
Smits H. (1985) *Financial Security as a Means to Improve Control of Transfrontier Movements of Hazardous Wastes,* OECD, Paris.
Stanners, D. and Bourdeau, P. (eds) (1995) *Europe's Environment: The Dobris Assessment,* European Environment Agency, Office for Official Publications of the European Communities, Luxembourg.
Starr, C. and Whipple, C. (1980) Risks of risk decisions. *Science,* **208**, 1114.
Stewart, R. (1993) Environmental regulation and international competitiveness. *Yale Law Journal,* **102**, 2039–2106.
Stoett, P. J. (1995) *Atoms, Whales, and Rivers: Global Environmental Security and International Organization,* Nova Science Publishers, Commack, New York.
Strohm, L.A. (1993) The environmental politics of the international waste trade. *Journal of Environment and Development,* **2** (2), 129–53.
Strong, M. (1993) Beyond Rio: prospects and portents. *Colorado Journal of International Environmental Law and Policy,* **4**, 21–3.
Suess, M.J. and Huismans, J.W. (eds) (1983) *Management of Hazardous Wastes: Policy Guidelines and Code of Practice,* WHO Regional Publication, European Series No. 14, World Health Organization, Regional Office for Europe, Copenhagen.
Sullivan, T.F.P. (ed.) (1995) *Environmental Law Handbook,* 13th edn, Government Institutes, Rockville, Md.
Talbot, E.O. and Craun, G.F. (eds) (1995) *Introduction to Environmental Epidemiology,* Lewis Publishers/CRC Press, Boca Raton, Fla.
Teuber, G., Farmer, L. and Murphy, D. (eds) (1994) *Environmental Law and Ecological Responsibility: The Concept and Practice of Ecological Self-Organization.* John Wiley, Chichester.
The Steering Committee (ed.) (1989) Taiwan 2000: Balancing Economic Growth and Environmental Protection. The Executive Report. Taipeh, Taiwan.
Thompson, J. (1988) Laying Africa waste. *New African (UK),* **252**, 35–7.
Tietenberg T. (1992) *Environmental and Natural Resource Economics,* 3rd edn, HarperCollins, New York.
Tillman, A., Baumann, M., Eriksson, H. and Rydberg, T. (1991) *Calculation of Environmental Load,* Chalmers Industriteknik, Gothenburg.

Tolba, M.K. (ed.) (1988) *Evolving Environmental Perceptions: From Stockholm to Nairobi,* United Nations Environment Programme, UNEP, Nairobi.
Tolba, M.K. and El-Kholy, O. (1992) *The World Environment, 1972–1992,* Chapman & Hall, London.
Tolentino, A.S., Brabante, A.T. and David, M.D. (1990) Toxic chemicals and hazardous waste management in the Philippines. *Waste Management & Research* **8** (2) 123–7.
UN (United Nations) (1995) *Recommendations on the Transport of Dangerous Goods,* 9th revised edn, United Nations, New York.
UNCED (1992) *Agenda 21,* UN Conference on Environment and Development, Conches, Switzerland.
UN Economic and Social Council (1985) *Transport of Dangerous Goods/Orange Book of the UN,* United Nations, Paris.
UNEP (United Nations Environment Programme) (1988) *The State of the Environment,* UNEP, Nairobi.
UNEP (United Nations Environment Programme) (1993) *Environmental Data Report 1993–94,* Blackwell, Oxford.
UNEP (United Nations Environment Programme) (1994a) *Basel Convention on the Control of Transboundary Movements of Hazardous Wastes and their Disposal, 1989, and Decisions adopted by the first /1992/ and the second /1994/ Meetings of the Conference of the Parties,* Basel Convention Series/SBC, No. 94/008, Geneva.
UNEP (United Nations Environment Programme) (1994b) Parties to the Basel Convention on the Control of Transboundary Movements of Hazardous Waste and their Disposal, 1994, Geneva.
UNEP (United Nations Environment Programme) (1995a) Third Meeting of the Conference of the Parties to the Basel Convention, Sept. 1995, Geneva.
UNEP (United Nations Environment Programme) (1995b) *UNEP's New Way Forward:Environmental Law and Sustainable Development,* United Nations Environment Programme, Nairobi.
UNESCO (1990) *Hydrology and Water Resources for Sustainable Development in a Changing Environment,* Paris.
United States Congress (1989a) *International Export of U.S. Waste – Hearing before a Subcommittee of the Committee on Government Operations, House of Representatives, One Hundredth Congress, Second Session, July 14, 1988,* US Government Printing Office, Washington, DC.
United States Congress (1989b) *Waste Export Control – Hearing before the Subcommittee on Transportation and Hazardous Materials of the Committee on Energy and Commerce, House of Representatives, One Hundred First Congress, First Session, on H.R. 2525, July 27, 1989,* US Government Printing Office, Washington, DC.
USEPA (United States Environmental Protection Agency) (1984) *Risk Assessment and Management: Framework for Decision Making,* EPA 600/9–85–002, Washington, DC.
USEPA (United States Environmental Protection Agency) (1988) *Guidance for Conducting Remedial Investigations and Feasibility Studies Under CERCLA,* EPA/540/G-89/004, OSWER Directive 9355.3–01. Office of Emergency and Remedial Response, Washington, DC.
USEPA (United States Environmental Protection Agency) (1989a) *Risk Assessment Guidance for Superfund. Volume I – Human Health Evaluation Manual (Part A),* EPA/540/1 89/002, Office of Emergency and Remedial Response, Washington, DC.

USEPA (United States Environmental Protection Agency) (1989b) *Risk Assessment Guidance for Superfund. Volume II – Environmental Evaluation Manual,* EPA/540/1–89/001, Office of Emergency and Remedial Response, Washington, DC.

USEPA (United States Environmental Protection Agency) (1990) *Follow-up on EPA's Program to Control Export of Hazardous Waste,* Office of the Inspector General, Washington, D.C.

USEPA (US Environmental Protection Agency) (1995) *Vendor Information System for Innovative Treatment Technologies,* EPA-542-C-95–001, Technology Innovation Office, US EPA, Washington, DC.

Uva, M.D. and Bloom, J. (1989) Exporting pollution – the international waste trade. *Environment,* **31** (5) (June), 4–5, 43–4.

Van Leeuwen, C.J. and Hermens, J.L.M. (eds) (1995) *Risk Assessment of Chemicals: An Introduction,* Kluwer Academic Publishers, Dordrecht.

Vesilind, P.A., Peirce, J.J. and Weiner, R.F. (1994) *Environmental Engineering,* 3rd edn, Butterworth-Heinemann, Newton, Mass.

Voigt, D. (1993). The Maquiladora problem in the age of NAFTA: where will we find solutions? *Minnesota Journal of Global Trade,* **2**, 323.

Wagner, T. (1994) *In Our Backyard: A Guide to Understanding Pollution and its Effects,* Van Nostrand, Reinhold, New York.

Walsh, M.T. (1992) The global trade in hazardous wastes: domestic and international attempts to cope with a growing crisis in waste management. *Catholic University Law Review,* **42**, 103–41.

Ward, H. (1994) Trade and environment in the round – and after. *Journal of Environmental Law,* **6**, 263–95.

Warhurst, A. (1994) Environmental Degradation from Mining and Mineral Processing in Developing Countries: Corporate Responses and National Policies, OECD, Development Centre Documents, Paris.

WCED (The World Commission on Environment and Development) (1987) *Our Common Future,* Oxford University Press, Oxford.

Weiss, E.B. (1993) International environmental law: contemporary issues and the emergence of a new world order, *Georgetown Law Journal,* **81**, 675–710.

WHO (World Health Organization) (1983) *Management of Hazardous Waste,* WHO Regional European Series No.14.

Wilson, David C. and Balkau, F. (1990) Adapting hazardous waste management to the needs of developing countries – An overview and guide to action. *Waste Management & Research,* **8** (2), 87–97.

Woodside, G. (1993) *Hazardous Materials and Hazardous Waste Management: A Technical Guide,* John Wiley, New York.

World Bank (1996a) *Global Economic Prospects and the Developing Countries,* World Bank, Washington, DC.

World Bank (1996b) *Social Indicators of Development.* Published for the World Bank by the Johns Hopkins University Press, Baltimore, Md.

World Bank and European Investment Bank (1990) *The Environmental Program for the Mediterranean – Preserving a Shared Heritage and Managing a Common Resource,* The International Bank for Reconstruction and Development/World Bank, Washington, DC.

Wynne, B. (1987) *Risk Management and Hazardous Wastes: Implementation and Dialecties of Credibility,* Springer-Verlag, Heidelberg.

Wynne, B. (1989) The toxic waste trade: international regulatory issues and options. *Third World Quarterly*, **11** (3), 120–47.

Yakowitz H. (1989a) Global hazardous transfers. *Environmental Science & Technology*, **23** (5), 510–11.

Yakowitz, H. (1989b) *Monitoring and Control of Transfrontier Movements of Hazardous Wastes: An International Overview*, Technical Paper W/0587M, OECD, Paris.

Yakowitz, H. (1995) *The OECD Control System for International Trade in Recoverable Wastes*, Waste Management and Recycling International, Paris.

Zaelke, D.J. (1993) Making trade and environmental policies mutually reinforcing: forging competitive sustainability. *Environmental Law*, **23**, 545.

Zirm, K.L. and Mayer, J. (eds) (1990) *The Management of Hazardous Substances in the Environment*, Elsevier Science Publishers, London.

Recommended journals

Archives of Environmental Contamination and Toxicology, Springer-Verlag, New York.

Arizona Law Review, The University of Arizona College of Law, Tucson, Ariz.

Arizona State Law Journal, College of Law, Arizona State University, Temple, Ariz.

Boston College Environmental Affairs Law Review, Boston College Law School, Newton Centre, Mass.

Clean Air (and Environmental Protection), National Society for Clean Air and Environmental Protection, Brighton.

Clean Air, The Journal of the Clean Air Society of Australia and New Zealand, Clean Air Society of Australia and New Zealand, Eastwood, NSW.

Colorado Journal of International Environmental Law and Policy, University Press of Colorado, Niwot, Colo.

Columbia Journal of Environmental Law, School of Law, Columbia University, New York.

Critical Reviews in Environmental Science and Technology, CRC Press, Boca Raton, Fla.

Ecological Applications: A Publication of the Ecological Society of America, Washington, DC.

Ecological Modelling: International Journal on Ecological Modelling and Systems Ecology, Elsevier Science B.V., Amsterdam.

Ecological Research, The Ecological Society of Japan, Blackwell Science (Australia), Victoria.

Ecology Law Quarterly, Boalt Hall School of Law, University of California, Berkeley, Calif.

Ecotoxicology and Environmental Safety, Academic Press, Orlando, Fla.

Emory Law Journal, Emory University, Atlanta, Ga.

Environmental & Engineering Geoscience, AEG, Texas A&M University, College Station, Tex.

Environmental Geochemistry and Health, Chapman & Hall, London.

Environmental Health Perspectives: Journal of the National Institute of Environmental Health Sciences, Research Triangle Park, NC.

Environmental Health Perspectives Supplements, National Institute of Environmental Health Sciences, Research Triangle Park, NC.

Environmental History, University Press of Kansas, Lawrence, Kan.

Environmental Impact Assessment Review, Elsevier Science, New York.

Environmental Law, Northwestern School of Law of Lewis & Clark College, Portland, Oreg.

Environmental Management, Springer-Verlag, New York.
Environmental Monitoring and Assessment: An International Journal, Kluwer Academic Publishers, Dordrecht.
Environmental Policy and Law, International Council of Environmental Law (ICEL), Bonn.
Environmental Politics, Frank Cass, London.
Environmental Science & Technology, American Chemical Society, Washington, DC.
EnvironMetrics, John Wiley, Northampton.
Harvard Journal of Law and Public Policy, Harvard Law School, Cambridge, Mass.
Hastings International and Comparative Law Review, University of California, Hastings College of the Law, San Francisco.
Hastings Law Journal, Hastings College of the Law, San Francisco.
Hazardous Waste & Hazardous Materials, Mary Ann Liebert, Inc., Publishers, Larchmont, NY.
Indoor Environment, KARGER, Rothenfluh, Switzerland.
International Journal of Environment and Pollution, Inderscience Enterprises Ltd/UNESCO, Geneva.
International Journal of Environmental Health Research, CARFAX Publishing, Oxfordshire.
International Review of Law and Economics, Elsevier Science, New York.
Journal of Clean Technology and Environmental Sciences, Princeton Scientific, Princeton, NJ.
Journal of Environmental Law, Oxford University Press, Oxford.
Journal of Environmental Science and Health, Marcel Dekker, Monticello, NY.
Journal of Environmental Science and Health, Part A: Environmental Science and Engineering and Toxic and Hazardous Substance Control, Marcel Dekker, Monticello, NY.
Journal of Environmental Science and Health, Part B: Pesticides, Food Contaminants, and Agricultural Wastes, Marcel Dekker, Monticello, NY.
Journal of Environmental Science and Health, Part C: Environmental Carcinogenesis & Ecotoxicology Reviews, Marcel Dekker, Monticello, NY.
Journal of Environmental Systems, Baywood Publishing Company, Inc., Amityville, NY.
Journal of Environmental Technology, Publications Division, Selper Ltd, London.
Journal of Hazardous Materials, Elsevier, Amsterdam.
Journal of Soil Contamination, Lewis Publishers/CRC Press, Boca Raton, Fla.
Journal of the Air & Waste Management Association, AWMA, Pittsburgh, Pa.
Journal of the Institution of Water and Environmental Management, IWEM, London.
Journal of Toxicology and Environmental Health, Taylor & Francis, London.
Journal of World Trade, Geneva.
Law and Policy, Blackwell Publishers, Oxford.
Marine Policy: The International Journal of Ocean Affairs, Pergamon/Elsevier Science, Exeter.
Restoration Ecology: The Journal of the Society for Ecological Restoration, Blackwell Science, Cambridge, Mass.
Risk Analysis: An International Journal, Plenum Press, New York.
The Environmentalist, Chapman & Hall, Hampshire.
The Environmental Professional, NAEP, Blackwell Science, Cambridge, Mass.
The George Washington Journal of International Law and Economics, George Washington University, Washington, DC

The Science of the Total Environment (An International Journal for Scientific Research into the Environment and its Relationship with Man), Elsevier Science B.V., Amsterdam.

Toxic Substance Mechanisms, Taylor & Francis, London.

The Harvard Environmental Law Review, Harvard Law School, Cambridge, Mass.

UCLA Journal of Environmental Law and Policy, UCLA School of Law, Los Angeles.

Waste Management, Pergamon/Elsevier Science, Oxford.

Waste Management & Research, ISWA, Copenhagen.

Water, Air & Soil Pollution: An International Journal of Environmental Pollution, Kluwer Academic Publishers, London.

Yale Law and Public Policy, Yale Law School, New Haven, Conn.

Index